Effects of Ionospheric Scattering on Very-Long-Distance Radio Communication

Effects of Ionospheric Scattering on Very-Long-Distance Radio Communication

H. A. WHALE

Director, Radio Research Centre
The University of Auckland
Auckland, New Zealand

℗ Springer Science+Business Media, LLC • 1969

Library of Congress Catalog Card Number 76-84765

ISBN 978-1-4899-6254-6   ISBN 978-1-4899-6545-5 (eBook)
DOI 10.1007/978-1-4899-6545-5

© 1969 Springer Science+Business Media New York

Originally published by Plenum Press in 1969.
Softcover reprint of the hardcover 1st edition 1969

PREFACE

In this work I have attempted to present a reasonably coherent account of many aspects of long distance communication by radio that have received little attention in the past. Investigation of these problems was undertaken since New Zealand, by reason of its location and its traditional ties with Europe as a market and a supplier has problems in the field of radio communication which are probably amongst the most difficult in the world. These were particularly troublesome during World War II and it was in an effort to solve these problems that the Seagrove Radio Research Station was set up by the New Zealand Department of Scientific and Industrial Research as a section of the Physics Department of the University of Auckland. This was largely due to the efforts of the New Zealand Post Office and New Zealand Broadcasting Corporation (encouraged by Professor Emeritus P.W. Burbidge of the Physics Department, University of Auckland) who were directly concerned with maintaining long-distance communications. For the last few years this organization has operated as a Department (the Radio Research Centre) of the University of Auckland. Since operations at the station began in 1951 the main interest has been in long-distance propagation and this, because of the nature of the problems which have arisen, has led to an allied interest in ionospheric irregularities.

Even in 1951 it was felt by many that the subject of ionospheric radio propagation had been "worked out". It certainly was not one of the fashionable subjects in physics at that time. It is now evident that the approach to most of the problems which were then recognised was one of producing more accurate solutions to more and more idealized cases. In the following pages it will be seen that, even for short distance radio propagation, there is a limit to the accuracy with which predictions can be made. This does not mean that there is anything basically wrong with our predicting techniques but rather that there are fundamentally unpredictable (or random) factors present. These assume greater importance as the propagation distance increases until, at distances greater than about 15000 kilometres, the essentially random factors become more or less dominant. From this point of view it can be said that the work presented here is an extension of that in the classical text books on radio propagation where essentially short distances are involved but goes a long way towards replacing the classical concepts where the propagation is over long distances.

I have tried to indicate, in this text, that although many questions have been answered, many new problems have been raised. In particular, of course, we have dealt with only one receiving location in most cases. The problem now is whether, with the increasingly widespread use of satellites for communication purposes, this type of investigation should be continued. The answer may be affirmative since, although satellite relay systems are ideal for fixed point communication, they do not as yet provide a solution to two-way communication between isolated people and peoples.

CONTENTS

CHAPTER 1

INTRODUCTION

In the relatively few years since 1904 when Marconi first demonstrated the feasibility of communicating over relatively long distances by means of radio waves, a tremendous volume of work on the subject of radio propagation has been published. Much of it has been concerned with obtaining more accurate agreement between theory and experiment in radio transmission *via* the ionosphere. As the distance over which such experiments are carried out increases, the whole problem becomes more and more complicated since there are always some ionospheric properties which are unknown somewhere along the transmission path. If the path length (i.e. transmission distance) is small, the lack of detailed information on the properties of the ionosphere over some portions of the path may not present a very serious obstacle to obtaining a reasonable estimate of the behaviour of radio waves in traversing it because some idea of the properties of the unknown regions may be obtained by interpolating between those regions where the characteristics of the ionosphere are known. The errors introduced by this procedure tend, however, to be cumulative in their effects on the calculated results so that very large errors may occur if too many interpolations are made. In general it may be said that the problem becomes intractable (even using modern electronic computers) if the path length involved is more than about 10,000 kilometres. It is under these circumstances that it must be recognized that an approach such as that presented here which is essentially statistical in nature and does not attempt to study detailed behaviour may still yield useful results.

The aim in this work has been to attempt to present an account of those general characteristics of long-distance radio propagation which can be deduced from what is essentially a study of the average (and hence statistical) behaviour. In other words, we are forced to accept the fact that we will probably never be able to *specify* the characteristics of the ionosphere along the whole of a particular propagation path even for a short period of time, let alone *predict* these detailed characteristics for future times. It is pointed out in chapter 3 that, although some of the major variations in ionospheric characteristics are predictable

1

(the average diurnal variations), the minor variations are
largely unpredictable i.e. they are more or less random in
occurrence. While these latter variations may be minor in
relative scale they may nevertheless be of paramount importance in determining some of the major propagation effects.
We are therefore led to making measurements on the signals
as actually received after traversing a long path and to
attempting to specify, on the basis of these measurements,
the limits of the variations which may occur in the characteristics of these signals.

The major part of the experimental results on which the
work is based were obtained in Auckland; since there appeared
to be very special problems associated with receiving BBC
transmissions in New Zealand, many of the results were obtained using the normal short-wave broadcasts of the BBC as
signals. There were considerable difficulties associated
with this approach since these transmissions were not generally of the most suitable form for the type of investigation
for which they were used. It will be seen that the basic
data which is required is, at least, the variation of the
signal strength and the direction of arrival of the signal
with the time of day, with the frequency and with the location
of the transmitting station. The difficulties introduced by
using the available standard short-wave broadcasts arise from
the fact that, in general, a particular frequency is only
transmitted for a limited period each day. Unfortunately,
the gaps can not usually be filled in by using other stations
since stations in the same area tend to broadcast on similar
frequencies at similar times of the day. Some special
transmissions (e.g. the pulse transmissions discussed in
Chapter 7) have been used in an endeavour to solve some particular problems but the great amount of data normally
required for statistical analysis has made this approach im-
practicable in the general case. It will, in fact, be found
that much of the following is based solely upon measurements
of the direction of arrival of those signals which were available.

The situation as far as further investigations are
concerned is, however, improving since, with the advent of
artificial satellites which transmit radio signals, we now
have a source that is standardized and that, in time,
moves over almost the whole of the earth's surface. It is
certain that many of the questions left unanswered here will
be solved in the near future as these transmissions become

more common. The major problem in utilizing satellites for
this type of investigation arises from the fact that, in
order that long-distance propagation around the earth may
occur, the satellite must usually be located at a height
which is below that of the maximum electron density of the
ionosphere. This generally means that the satellite must
orbit the earth at a height of about 100 miles or less and
unfortunately the satellite then has a life-time of only a
few weeks so that many consecutive launchings of similar
satellites may be required in a long-term investigation.

CHAPTER 2

IONOSPHERIC PROPAGATION IN THE IDEAL CASE

The fundamental phenomena of ionospheric propagation
over short distances are well known and are discussed, for
example, by MITRA (1947)*. A most comprehensive handbook on
the subject of ionospheric radio propagation is that of
DAVIES (1965). The main relevant features in a simplified
typical case of ionospheric propagation are shown in fig.
2.1. Here we have a transmitter located on the earth's surface at T and we will consider the fate of the four emitted
rays T1, T2, T3 and T4, ignoring the fact that, because of
the earth's magnetic field, each ray is generally split into
two magneto-ionic components (the ordinary and extraordinary
rays) which travel along somewhat different paths.

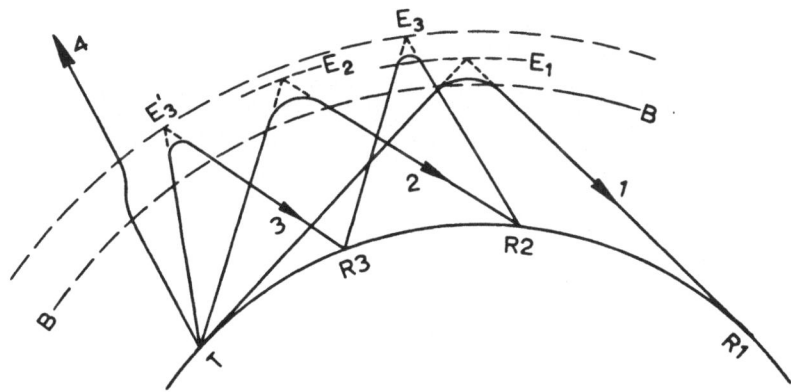

*Fig. 2.1. Idealized hop-type ionospheric propagation showing
the different effective reflection heights for different
angles of incidence of the rays.*

---

*All references are given after Chapter 21.

Ray 1 is the ray at the lowest possible angle of elevation i.e. it is tangential to the earth's surface at the transmitter at T. When this ray encounters the ionosphere (whose base is at B) it is refracted so that it returns to the earth, striking it again tangentially at R1. A great simplification is introduced if, instead of considering the actual curved ray path in the ionosphere, we regard the ray as travelling in straight lines and as being returned from a perfectly reflecting layer at E_1. Then the height of F_1 above the surface of the earth is called the "effective" or "virtual" height of the reflecting layer for this particular ray. We note that, in this particular case, even a small tilt of the layer near to E_1 can cause ray 1 either to miss the earth completely or to strike it at an elevation angle which is substantially greater than the tangential angle of zero degrees.

Ray 2 is typical of single hop transmission between T and a receiver placed at R2. The height of E_2 above the earth's surface (i.e. the effective height of the reflection point) is somewhat greater than that of E_1. Similarly, ray 3 leaving the transmitter at a slightly greater elevation angle will be reflected at a slightly greater effective height and will reach the ground at R3, say. It will then be reflected from the ground, will make another encounter with the ionosphere and, if the distance TR3 is exactly half TR2, it will reach the ground again at R2. This is the case of a two-hop transmission mode.

If the ionosphere is sufficiently dense, i.e. if the maximum electron density is sufficiently high, multiple hop modes of very much higher order may be possible.

As a final basic type of propagation, we have the ray T4 which is incident on the ionosphere at such a high elevation angle that it passes right through and will, in general, be lost in space. For this to occur, the transmitted frequency must be greater than the *critical* frequency f_c of the ionosphere. This critical frequency is determined by the maximum electron density of the ionosphere; it is one of the quantities which are measured regularly by the world-wide network of ionospheric sounding (ionosonde) stations and which are the subject of both short and long-term predictions by various government agencies. The best-known of these are the Radio Research Station, Slough, the National Bureau of Standards, Washington and the Ionospheric Prediction Service, Sydney.

In the very simple examples which we have considered above, none of the paths is more complicated than a two-hop mode. However, in order that exact predictions for the characteristics of even these simple paths may be made, it is necessary to have detailed knowledge of the properties of the ionosphere in the general regions near E_1, E_3, E_2 and between E_3' and a point directly above the transmitter. Such complete information is seldom available although some special measurements have been made by various workers under conditions which approached the ideal (i.e. all the relevant properties of the ionosphere were measured at the time of transmission). It is not our purpose to investigate or discuss the use of predicted values of ionospheric parameters in propagation problems since this subject is already very adequately covered in the literature. We will, however, consider (in Chapter 3) measurements made on a relatively simple radio link with the aim of estimating the accuracy and usefulness of such predictions in practice. It can be seen from fig. 2.1 that the propagation mode can be identified if measurements of the elevation angle are made at the receiving site. This is, in itself, a fairly difficult measurement to make but some of the methods which have been used are described in Chapter 4, for example. In practice, measurements of both elevation angle and bearing angle are made and these are generally referred to jointly as the *direction of arrival*. As the ionosphere changes through the day, it is to be expected that the elevation angle, in particular, will show some more or less regular variation. It is, in fact, found that although the averaged diurnal variations of the directions of arrival of the signals show a well-defined behaviour (when the averaging is carried out over about 30 days) there are significant departures from this average on any particular day. This is not to be taken as indicating that the predictions of the characteristics of the propagation path which are themselves based on predicted general ionospheric characteristics are of no value. Rather, it is confirmation of the generally accepted rule that ionospheric changes occurring in a time of less than about one month are essentially random and hence largely unpredictable.

CHAPTER 3

MEASUREMENTS ON A SHORT PATH

From a receiving station at R (in fig. 2.1) it should be possible, as has been pointed out, to determine the propagation mode (whether one-hop or two-hop, for example) by taking measurements on the incoming wave. As an example of the complications which occur in practice on even such a simple communications path we will consider the results of measurements taken at Auckland on transmissions from Brisbane, a distance of 2,250 kilometres. These measurements of the elevation and bearing angles were made on a rotating interferometer located at Seagrove, near Auckland. This instrument is described in the next chapter; at this stage it is sufficient to mention that, being a phase-comparison device, it yields a result which refers to the strongest wave in a group of incoming waves.

Firstly, it is found that, with any ionospherically reflected wave, the received signal *scintillates*. This is a term borrowed from the astronomers who use it to describe the fluctuations in brightness of a star which in their case arise from variations in the refractive index of the atmosphere and are an indication of atmospheric turbulence. They are accompanied by small fluctuations in the apparent position of the star. Similarly, in the case of radio waves, it is found that the received signal varies in its amplitude or "brightness" (i.e. it "fades") with typical fading periods of a few seconds and, at the same time, the apparent incoming direction or "position" also fluctuates with about the same period. This *radio* scintillation arises, not directly from variations in atmospheric density, but from variations in the electron density of the ionosphere. These latter variations do, however, seem to arise largely as a result of variations in the density of the atmosphere in which the ionosphere is "imbedded". In fact, of course, the ionosphere and its associated neutral atmosphere are very closely tied together since any change in electron density must be accompanied by a change in positive ion density, the medium as a whole remaining neutral; these large positive ions will then communicate any velocity changes to the neutral particles (also large compared with electrons) so that the atmosphere as a whole will tend to hang together.

9

We may note here that the fluctuations in the strength
of the radio signal tend to be related to the fluctuations
in the incoming direction of the wave, the weak signals
tending to occur at times when the deviation of the wave
from its mean position is greatest. This effect is discussed in more detail in Chapter 9 although a preliminary
discussion of a simple case in which it occurs is given
below in this chapter. For measurement purposes we often
regard these scintillations as a complicating factor which
could, with advantage, be removed. This can be achieved by
an averaging process; in the measurements to be described
in the section, the averaging time was about half an hour so
that the effects of scintillations (which occupy times of the
order of a few seconds) were completely removed.

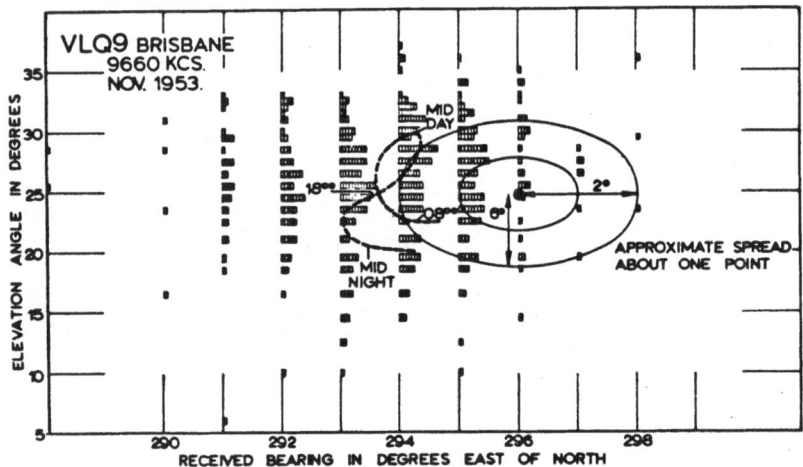

Fig. 3.1   The three types of variation in the received
           angle of arrival.
           Dashed line - the diurnal variation averaged over
               the whole month.
           Small rectangles - the actual measured directions
               averaged over one hour to remove the scintilla-
               tion effects.
           Ellipses - the approximate spread of incoming
               directions arising from scintillation effects.

If these *averaged* measurements of bearing and elevation angle are repeated on the same station on a series of successive days, it is found that two effects appear. Firstly, there is some similarity between the measurements obtained on different days and these similarities persist over periods of a few weeks. Secondly, the differences between particular readings obtained at a particular time of day and the mean obtained by taking long-term averages (for this particular time of day) over about a month appear to be random.

This behaviour is illustrated in fig. 3.1 in which the average diurnal variation of elevation angle and bearing angle for the month of November 1963 is plotted as the heavy dashed curve and the individual daily measured values are also shown. It has been found that the averaged results agree very well with what would be calculated from the averaged ionospheric characteristics for that particular month. On the other hand, the day-to-day variations from this average are largely random and, as such, unpredictable. Many more examples of the way in which the elevation angle and the bearing angle vary through the day are given in a paper (WHALE, 1956) which discusses the use of these curves for determining how much the ionosphere is tilted. It will be noted that there are three kinds of variation depicted in fig. 3.1. Firstly, there is the average diurnal variation (the heavy dashed line) which can be expected to be predictable from the average ionospheric characteristics over the transmission path. Note the difference in scale in the vertical and horizontal axes. It is normal, in such a short-range transmission, for the variations in elevation angle to be very much greater than those in the bearing angle. Secondly, there are the day-to-day variations from this average, exemplified by the small rectangles, each of which represents results averaged over about one hour so that the scintillations have been removed. Thirdly, there is the scintillation effect - a typical spread of the apparent received direction of the scintillating wave is indicated by the ellipses shown. On this particular path it is normal for the scintillation in the vertical direction to exceed that in the horizontal direction. Although the ellipses in fig. 3.1 are drawn with their axes vertical and horizontal, this is not a necessary, nor perhaps even a typical, orientation. The size and orientation of the scintillation ellipse may be measured by methods which are discussed in Chapter 15 where some examples of the results to be expected are presented. A simplified discussion of the way in which scintillation effects arise from the interference of two waves is presented later in this section.

These three types of fluctuation are differentiated mainly by the time scales involved. The diurnal variations occur with a period of 24 hours, the day-to-day fluctuations with periods of a few hours and the scintillations with periods of less than about a minute. If we regard the ionosphere as moving overhead then all three types of variation could be regarded as arising from irregularities occurring in a postulated uniform ionosphere. For the diurnal variations these irregularities could be regarded as being of very large size and moving around the earth once in 24 hours. For the day-to-day variations we are concerned with ionospheric structure which is one or two orders of magnitude smaller than this and for the scintillations we are concerned with irregularities which may be of the order of a few hundreds of metres or less in size. This difference in scale of the irregularities is also associated with the major difference between the day-to-day variations and the scintillations i.e. in the former case we are concerned with variations in the mean direction of the wave while in the case of scintillations we are concerned with the effects produced by several waves arriving at the receiver simultaneously. We shall see later that this is the same as the distinction between reception in the *near zone* (where it is the refraction of one main ray which is important) and in the *far zone* (where interference effects occur) of a diffracting screen.

The above example refers to a fairly short path, i.e. a little longer than 2000 kms. As the distance between the transmitter and the receiver increases there is still a recognizable average diurnal variation but the day-to-day departures from this average tend to increase. Since the number of possible modes (i.e. paths with different numbers of hops between the transmitter and receiver) increases rather rapidly with distance and since each mode is associated with a different elevation angle, it becomes meaningless to speak of the "average" elevation angle at the greater distances. However, the bearing of the incoming wave will not, in general, change greatly with the number of hops, except when there are considerable transverse tilts in the ionosphere. In practically every case, the changes in bearing angle with number of hops will generally be small compared with the changes in elevation angle. The absolute limit to the range of angles covered by the elevation angle is, of course, from zero (glancing incidence) to 90° (vertical incidence). Even at transmitter-to-receiver distances

which approach antipodal (i.e. the transmitter and receiver
are both on a diameter through the centre of the earth) the
day-to-day deviations of the bearing from the mean bearing
(averaged over about a month) will seldom approach ±45°.
The mean bearing may itself change by 180° during the day,
but it is the deviations from this with which we are concerned here. Although the elevation angle measurements may
be essential in determining modes of propagation it is the
day-to-day variations of the bearing angle from the mean
bearing which yield the most information on the propagation
processes over the greater distances. Such variations have
been obtained experimentally for stations at varying distances, the results being discussed more fully in Chapter 7.

We have used the term "bearing angle" in the above with
no exact definition of what this means. For the case of a
plane wave travelling over a uniform smooth ground (as is
the case, for example, with the ground wave from a medium
wave broadcasting station when there are no ionospheric
reflections of importance) the meaning of "bearing" is
obvious. It is the azimuthal angle measured to the East
from true North, of the direction from which the energy in
the wave is arriving. In this case, the wave-front is a
plane surface (by definition of a plane wave) and the wave-
normal (the forward perpendicular to the wave-front) is in
the direction of the energy flow.

When, however, the wave which is being received has
encountered some type of irregularities in its passage from
the transmitter to the receiver, the case is very different
from the above. In general, a collection of waves will be
received simultaneously. Some of these will contain considerable energy and others may be very weak; in some cases,
it may occur that the ratio of energy between the various
waves is changing continually. Under these circumstances it
is still usually possible to define a direction for the mean
or average power flow; in the case where the irregularities
are in the ionosphere an averaging time greater than about
five minutes is usually sufficient to remove most of the
fluctuations in the direction of the energy flow. The disadvantage of such an averaging technique is, of course, that
a large amount of significant information about the structure
of the incoming ensemble of waves is thereby lost and, in
very many cases, it is this information that is required. It
is then necessary to measure and record some quantity which
is related to the *instantaneous* wave-normal direction of the

composite wave which results. The fluctuation in wave-normal
direction which occurs when a wave is built up of a number of
independent waves, is, as has been pointed out above, often
called scintillation. The way in which this scintillation
arises may be seen from the following.

Consider the two plane waves labelled 1 and α in fig.
3.2 (a). These are represented by their wave normal directions and we will let them have amplitudes of 1 and α respectively. In order to provide an idea of the size of the
interference pattern produced when these two waves exist
simultaneously, we will specify that the angle θ between them
is 10° and let their wavelengths have the same value of 30m.
The resulting wavefronts can now be calculated; some representative results are drawn in fig. 3.2. Five distinct conditions illustrate the type of effects which can be expected.

(i) α = 0 : In this case only one wave is present
and thus the wavefronts are parallel lines which are perpendicular to the wave direction of wave 1. (Figure 3.2(b)).

(ii) α = ∞ : This is the case where the amplitude of
wave 1 is negligible compared with the amplitude of wave α.
The resulting wavefronts are a series of parallel lines all
perpendicular to the wave direction of wave α. (Fig. 3.2(f)).

(iii) α = .9 : The resulting wavefronts are drawn as
full lines. (Fig. 3.2(c)). If one wavefront is followed
laterally and the dotted lines drawn as averages through the
wavefront, it is seen that these "average" wavefronts are
parallel to the wavefronts for α = 0.

(iv) α = 1.1: This is obtained by reversing the roles
which the waves 1 and α played in producing the curves in
(iii) above. We now see that the "average" wavefronts are
parallel to those for α = ∞. (Fig. 3.2(e)).

(v) α = 1.0: This is a very special critical case.
The wavefronts are now discontinuous and are, in fact,
undefined at the discontinuities since, under this special
condition the resultant field is zero at these places.
These curves represent the changeover condition from α < 1
to α > 1 i.e. from (iii) to (iv). (Fig. 3.2(d)).

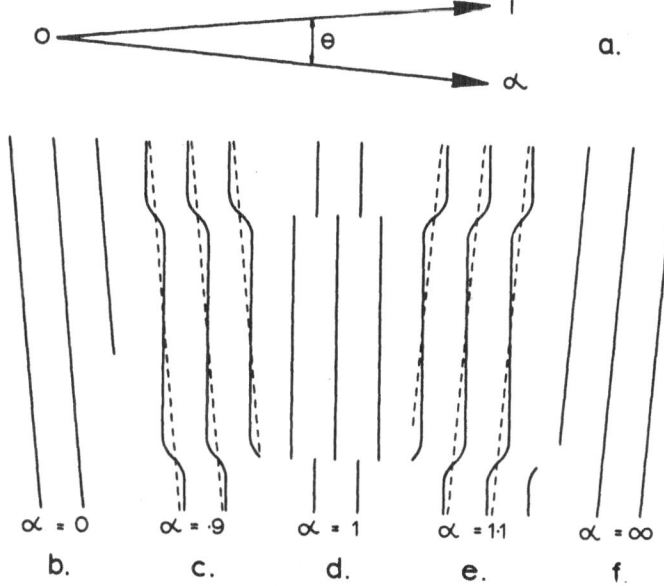

Fig. 3.2 *Resultant wavefronts produced by two interfering
waves. The mean wavefronts (dashed lines) are
controlled by the direction of the wave α when
α > 1 and by the wave 1 when α < 1.*

As a general conclusion it may be stated that, when
there are two dominant waves present, the resulting "average"
wavefronts follow those of the wave with the greater amplitude. The bends or kinks in the wavefronts become more
marked as the amplitudes of the two waves become more nearly
equal until, when they are exactly equal, sharp discontinuities appear.

Let us now investigate the problem of making accurate
measurements of the shape of these resultant wavefronts.
The patterns drawn in fig. 3.2 are frozen i.e. they refer to
a particular instant of time; they would be, in an actual
case, moving across the page at a velocity slightly in
excess of the free space velocity of a radio wave. If the
form of the pattern does not change with time, (i.e. the
phase relationship between the two component waves is fixed)
then it could, theoretically, be traced out in one of two

ways. Either some type of probe could be moved about and the
phase pattern mapped out or else a very large number of
stationary probes could be employed and the phase pattern
deduced from the resulting array of phase measurements. One
point which should be stressed here is that the actual
"phase" of a wave can not usually be measured in practice.
The measurable quantity is the difference in phase between
two parts of the wave or between a signal derived from the
wave and a reference signal which has *exactly* the same frequency. This means that, in either of the above methods of
investigation, the actual measurements would consist of the
phase differences between a fixed probe and either the moving
probe or each of the array of probes. Since the scale of the
pattern may be very large (e.g. 200 metres or more even in
the case considered in fig. 3.2) the practical difficulties
could be very great.

     In a practical case, when we are studying actual radio
waves which have been reflected from the ionosphere, the
phase difference between two component waves such as those
considered above will not be fixed. Usually, arising from
the diurnal changes in ionospheric structure, one wave will
be found to be slipping behind the other because the rate of
change in the path length is different for each of the two
rays. This has the effect of introducing a lateral shift
into the phase interference pattern (the wave-front pattern
which we are attempting to measure) so that the whole pattern drifts across a measuring location. Under these circumstances, the measurement becomes much simpler because we
can now take measurements over a small area and let the
pattern drift over the observing point. An example of a
result obtained by this technique is given later in Chapter
14.

     In the next few chapters we will be concerned more with
the information which can be obtained from measurements of
the bearing angle of waves which have travelled over different
distances than with information from elevation angle measurements. Some of the results which will be presented are derived from studies of the day-to-day fluctuations of the
bearing angle from its mean value; some of the further results depend on measurements of the spread of the incoming
cone of rays i.e. the degree of angular scintillation of the
wave. In both cases, some instrument which will carry out
the required measurements is needed. A particular type of
instrument, called a rotating interferometer (described in

in Chapter 4), has been found to be exceptionally versatile in that it can be employed for almost all of these measurements. Various other instruments which have been used for some special measurements, particularly in the study of scintillations, will be described in conjunction with a discussion of these other results in later chapters.

CHAPTER 4

MEASUREMENT OF ANGLE-OF-ARRIVAL

While a large number of different types of
direction-finder for obtaining the *bearings* of radio
signals have been developed, ranging from simple vertical loop antennas to very complicated systems such as
the Wullenweber type in which a narrow beam is used to
scan the incoming wave system, very few of them can be
applied to the problem of measuring vertical angles as
well. If the bearing of a signal is known, then the
problem of measuring the elevation angle δ reduces to
the problem of measuring the phase difference between
two antennas A,B spaced along the bearing line as in
fig. 4.1(A). The fact that there is also a wave which
is reflected from the ground arriving at each antenna
does not alter the phase difference between them
provided that the ground is flat and that its reflecting
properties near A are the same as those near B. It is
assumed that the antennas A and B are identical; in
particular, that they are at the same height above the
plane surface of the ground. A surface is plane, in the
radio sense, if it contains no irregularities approaching a quarter wavelength in size so that, at 30 MHz
(i.e. 10 metres wavelength) a flat field is a very plane
surface. This measurement of the phase difference
between the signals arriving at A and B may be thought
of as determining the distance AB' since the phase
difference between A and B is the same as the phase
difference between A and B'. Then the elevation angle
is derived from the simple relationship

$$\cos\delta = AB'/AB.$$

If the bearing angle is not known, both bearing and
elevation angles may be found if two measurements similar to the preceding are made. In this case the
antennas could be arranged at the corners of a triangle
as in fig. 4.1(B). Then, from measurements of the phase
difference between any two pairs of these antennas it is
possible to calculate both the bearing and elevation
angles.

From a practical point of view, such measurements
are difficult since, for accuracy, either a calibration

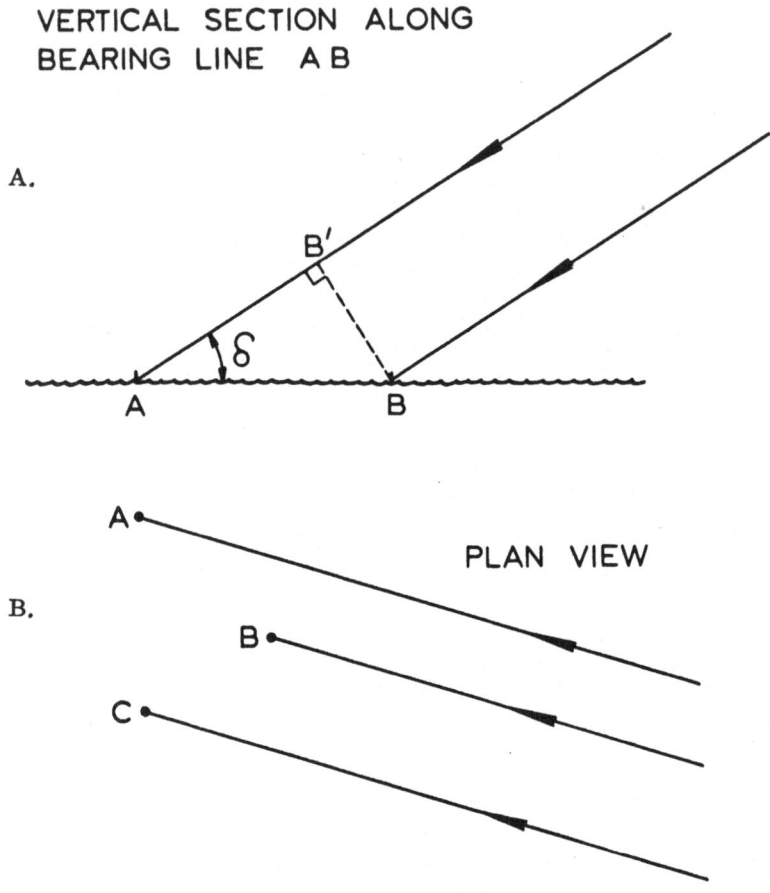

VERTICAL SECTION ALONG
BEARING LINE A B

A.

PLAN VIEW

B.

Fig. 4.1   A.   *The vertical angle δ can be obtained if the*
                phase-difference between two points A and B
                along a bearing line is measured.

           B.   *Both the vertical angle and the bearing angle*
                can be obtained if a three-antenna system is
                employed.

procedure is necessary for each frequency at which the
system is employed or else at least three identical phase-
stable antenna, transmission-line, amplifier systems must
be employed. It is true that such systems are employed
successfully in many radio-astronomy systems but these
usually have taken a considerable time to set up accurately and are seldom, if ever, used to cover a wide range
of frequencies.

The difficulty of building and maintaining accurately
phase-matched antennas and amplifiers can be overcome by
using a moving antenna system. It is, in practice,
relatively simple to determine when the phase difference
between two signals is 90°. Imagine, in figure 4.1(A),
that the antenna A is fixed in position and antenna B
is moved along the bearing line continuously and its
position marked when the phase difference between the
signals A and B is $\pm 90^{\circ}$ (i.e. the phase difference is 90°
or 270°). In this way, half-wavelengths may be stepped
off. But these are half-waves measured along the ground
and will thus be longer than the free-space half-waves in
the ration AB/AB'. The relationship is, of course,

$$\cos\delta = \lambda/\lambda g$$
where λ = free space wave-length,
λg = ground wave-length just measured.

A simple modification of the system allows the bearing
angle to be measured as well. In this system, the moving
antenna B would travel in a circle of a few wavelengths
diameter with the fixed antenna A somewhere inside the
circle. For simplicity, the fixed antenna can be placed
at the centre of the circle. Then, although the actual
physical distance between A and B remains constant, the
separation when measured in a particular direction (say
along a North-South line) will vary in a simple harmonic
manner as B revolves around A. The separation will also
vary in a similar way along another direction (say an East-
West line) but with a time displacement from the first
variation. In this case, the East-West separation is a
maximum when the North-South separation is a minimum
(zero) and vice versa. The system can thus be regarded
as being equivalent to two systems of antenna pairs with
variable spacings set up at right angles. With such a
system both elevation angle and bearing angle can be
measured.

B   A

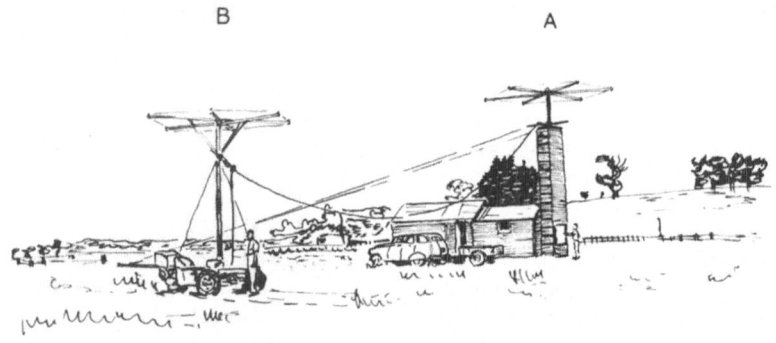

SEAGROVE — ROTATING INTERFEROMETER — 1953-1960

Fig. 4.2 The first rotating interferometer. The distance AB between the two antennas was about 60 metres.

        The first rotating interferometer, which operated exactly in this manner, is shown in fig. 4.2. The moving antenna B was mounted on a small electrically driven wagon which moved in a circular path around the fixed antenna A. The track radius was about 60 metres. Suppose that equipment which can determine when the phase-difference between the signals from the two antennas is 90° is housed at A. Then, since the success of the system depends on accurate location of the positions of B where this condition occurs it is necessary that B's position be accurately recorded at A. It is also necessary that B follow a closely circular path and of course that the signal received on antenna B be transferred to A for comparison. Since we are only interested in the way in which the phase difference between A and B changes as B revolves and thus do not require to know the absolute value of this phase difference at any time, we do not need to know the phase delay introduced by any amplifiers at B or by the transmission line connecting B to the measuring equipment at A. A simple method of accomplishing this is shown in fig. 4.3. The truck is provided with a steering arm extending forward as shown. This arm is "tethered" to a central mast which can rotate, and, in doing so, drive a recorder drum in synchronism with the truck (one rotation of the drum per lap of the truck). The steering wheel of the truck is spring-loaded so that, when driving forwards, the truck will tend to run in a spiral of increasing radius. When the tethering wires tighten, the steering arm will correct this tendency and the track followed will approach a circle, the error decreasing exponentially with distance covered.

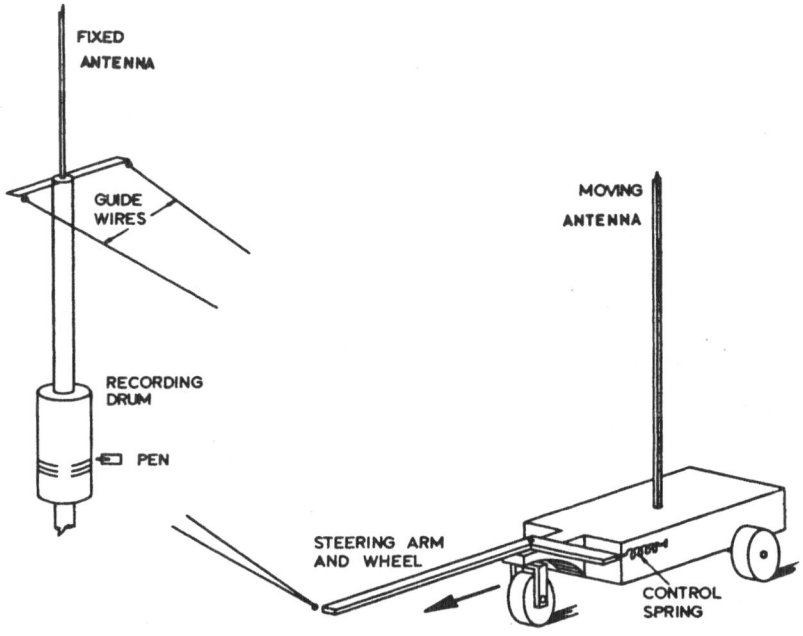

Fig. 4.3 The mechanical system employed in the rotating interferometer.

Represent the signal from the A-antenna as a vector
(A in fig. 4.4) and that from the B-antenna as the vector
B. Then, we assume that we have available equipment which
can indicate when the vector B is at B' or B" i.e. when the
phase difference is 90°. The position of the vector B
with respect to A will depend on the relative positions
of the two antennas; in fact, if we have the situation
drawn in figure 4.5 (a plan view of the rotating interferometer) where the vector B coincides in direction with the
vector A when the truck is at the positions A', then
vector B will be in the directions B', B" (in fig. 4.4)
when the truck is at the positions B', B" respectively
in fig. 4.5. It is easily arranged that, as a recording
system, a pen marks the rotating chart drum when B is in
the half-plane B'CB" and leaves a blank when B is in the
half-plane B"AB' (in fig. 4.4). If successive rotations
are recorded on adjacent lines then a typical record will
be as shown in figure 4.6.

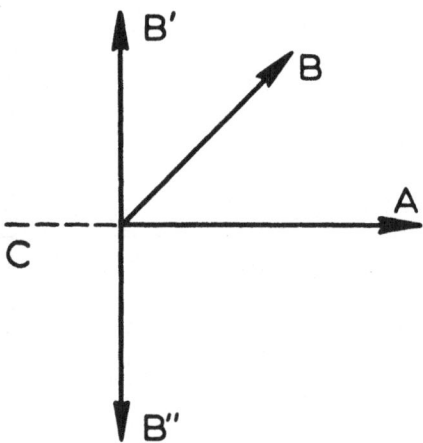

*Fig. 4.4   Vector diagram of the signals from the antennas
at A and B.*

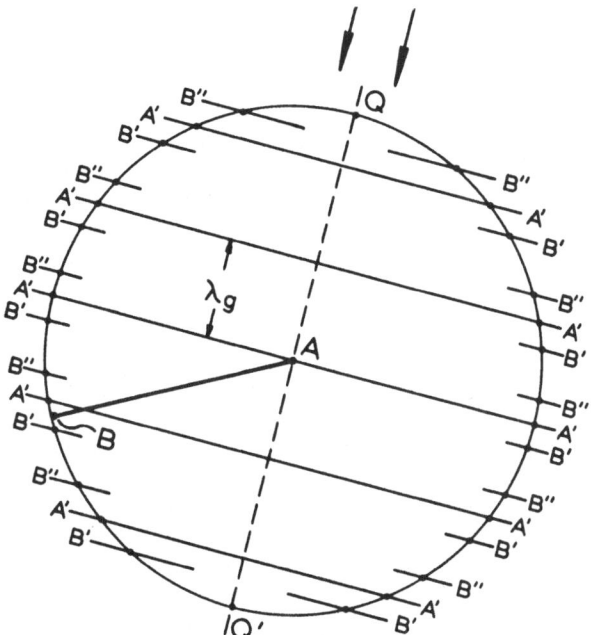

Fig. 4.5 *Plan view of the path of the antennas in a rotating
           interferometer. The signal at the antenna B is
           in phase with that at A when B occupies the posi-
           tions A'*

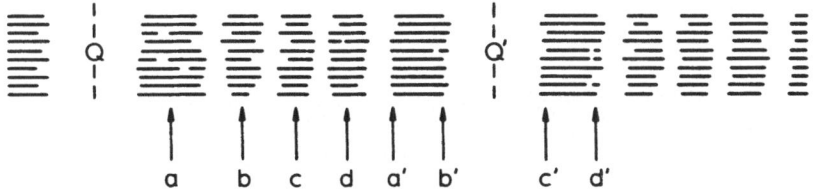

Fig. 4.6 *The type of record obtained with a rotating inter-
           ferometer. The chart has been unwrapped from the
           drum shown in figure 4.3. The uneven nature of
           the edges of the blocks in the pattern arises from
           scintillation effects and may be used to measure
           such effects.*

     In order to obtain bearings and elevation angles
from this record i.e. to analyse it, we first note that it
is symmetrical about the line QQ'. This line is easily
found and gives the bearing of the wave. We next note
that the number of bands or fringes between Q and Q' is
related to the wavelength along the ground and hence
to the elevation angle. The details of one method of
analysis, which is quite straightforward once the appropriate scale (shown at the lower edge of fig. 4.10) has
been constructed, are given in a paper by WHALE and
BANNISTER (1966).

     The type of construction described above which requires
a truck and a roadway for the moving aerial is very suitable
where large spacings between the antennas are required.
This is the case where the frequency to be investigated
is low or where the elevation angles to be measured are
very large. The system just described, where the spacing
was about 60m, worked satisfactorily at frequencies above
5 MHz at all elevation angles normally encountered in
short-wave radio propagation i.e. up to about 40°.

SEAGROVE — 1962

*Fig. 4.7 Elevated beam rotating interferometer. In this
example, the antenna spacing is 40 metres.*

The great disadvantage of the system arose in fact from
the excellent control of the track followed by the moving
wagon provided by the steering mechanism described above.
Since the wheels followed almost exactly the same path
(within about $\frac{1}{4}$" on most occasions) on each orbit (as it
were) the roadway was subjected to repetitive loads in a
manner which is not encountered on highways, for example.
Depending on the type of ground on which the track was
laid, this could (and, in fact, did) lead to very difficult
maintenance problems.

It is possible, of course, to realize the same relative
positioning of the antennas in another way. If the two
antennas A and B are mounted on the ends of a boom which
is pivoted in the centre, (as in fig. 4.7) it can be seen
that one rotation of the boom is equivalent to one circuit
of the track. Such a system is easily realized for moderate
antenna spacings, the rotating interferometer shown in
figure 4.7 having, for example, an antenna spacing of 40
metres.

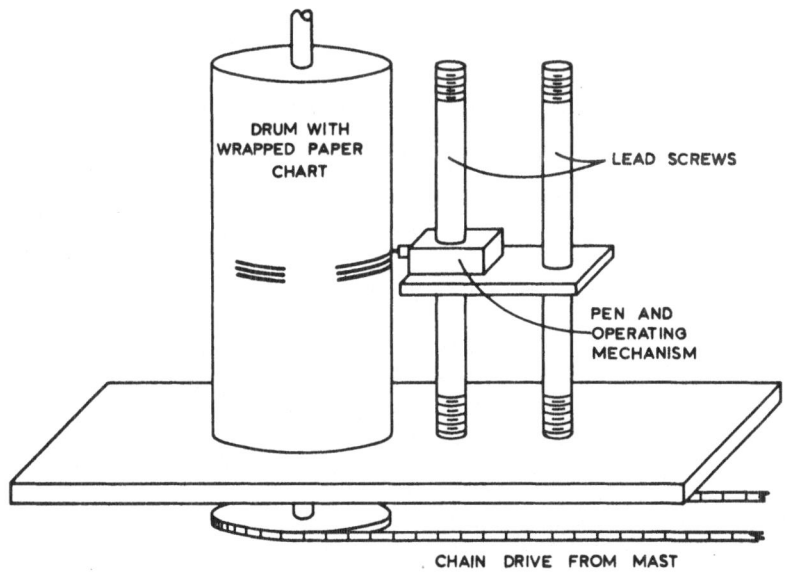

*Fig. 4.8 Vertical drum type recording system. The chart
length is determined by the size of the drum and
hence charts must be renewed frequently.*

In many investigations of short-wave propagation phenomena, speed in obtaining results is not of prime importance since changes occur at relatively slow rates.   In
general, a moderately sized interferometer would be
operated at a rotation speed of about once per few minutes.
However, as will be discussed later, some investigations
can best be carried out using satellites as signal sources.
In this case it is essential to obtain results more rapidly.
A small rotating interferometer, which was specifically
designed to operate on satellite signals at frequencies
above about 10 MHz, with an antenna spacing of 20 metres
and which rotates once every 8 seconds has been employed
to obtain some such results.

A simple recording system would use a paper chart
wrapped around a drum which rotates with the interferometer
(fig. 4.8).   Each line of recording is spaced from the
preceding one - a reasonable spacing is 40 lines per inch.
At 5 minutes per revolution, 24 hours of recording can be
accommodated on a drum which is 8" high, so that the charts
need to be renewed only once a day.   On the other hand,
with 8 seconds per revolution, paper is used at the rate
of 22'6" per day so that a continuous system of paper

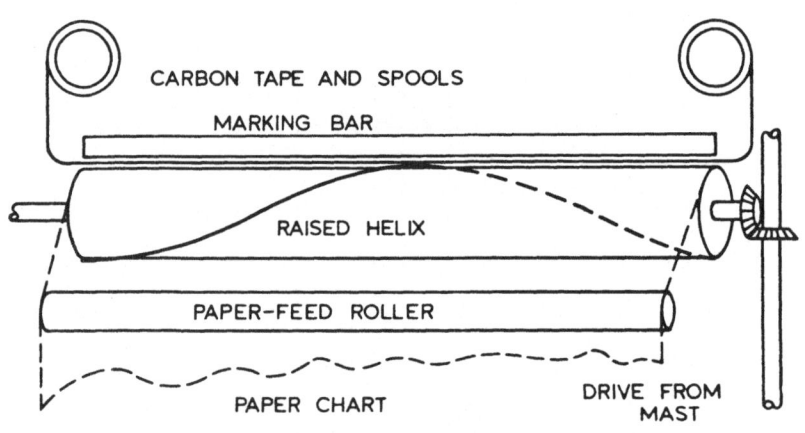

*Line 4.9 Horizontal drum type recording system. The chart
need only be renewed at long intervals.*

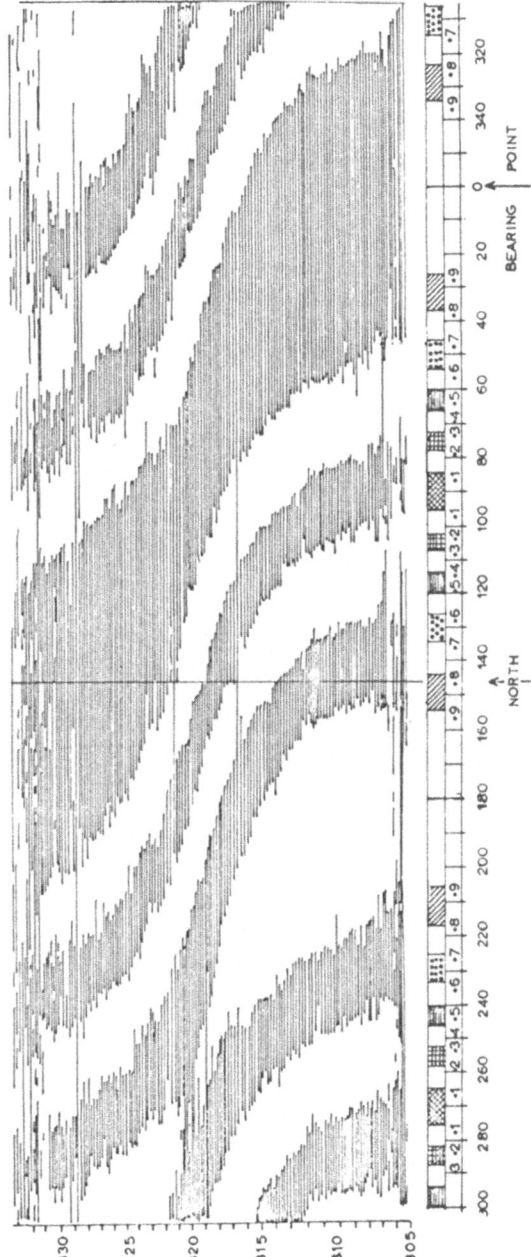

Fig. 4.10 Typical record obtained with the fast interferometer using a satellite as the source of the signal. The scale used in the analysis of the records is shown at the bottom of the chart.

feed is required. Such a system is shown in figure 4.9
where a roll of paper is fed over a horizontal drum with
a raised spiral on its surface. This drum rotates with
the mast and the record is obtained by pressing a type-
writer ribbon on to the paper with a marker bar which is
solenoid-operated. A typical record of transmissions
from one of the Beacon S66 series of satellites is shown
in figure 4.10. It may be mentioned here that, as can
be seen immediately from this record, a satellite produces
a very distinctive type of pattern. This arises, of
course, from the fact that the satellite is moving very
rapidly compared with any ground-based transmitters or
any natural or man-made sources of interference. It has
been found that the distinctive nature of the records can
be extremely useful in identifying those times when signals
from satellites are being received.

CHAPTER 5

SPHERICAL EARTH PROPAGATION - THE IDEALIZED CASE

Before describing some of the experimental findings on
long distance propagation paths, we will consider an idealized situation. This is the case of a smooth spherical earth
surrounded by a smooth spherical reflecting ionospheric shell
which is concentric with the earth. A ray emitted by a
transmitter will travel by a series of hops between the earth
and the ionosphere. Since we have assumed that both the
earth and the ionosphere are smooth and that they are concentric, such a path will always lie in some plane which
passes through the centre of the earth. The intersection of
this plane with the surface of the earth is a *great circle*
(by definition). By contrast, the intersection of a plane
which does not pass through the centre of the earth with the
surface of the earth is called a *small circle*. We need not
consider the actual hops at this stage but merely note that
the projection of the path on to the surface of the earth
(the "trace" of the path) is a great circle.

All great circles from the transmitter at T (in fig.
5.1) pass through the *antipodal point* (AP) and there is only
one great circle which passes through T and any other point
R which is not AP. There are two paths from T to R; one of
these, which does not go through AP is called the *short path*
and the other is called the *long path*. Any point on the
surface can be reached by following some great circle which
goes through T i.e. a transmitter at T would be received
everywhere on the earth in this idealized case.

There are a number of ways in which maps are designed
so that it is particularly easy to draw great circles on
them. This problem is, of course, a long-standing one since
a great circle is the shortest distance on the surface of
the earth between two points and is thus usually the quickest
route for a ship or an aircraft to follow. The most common
map of the world is drawn on Mercator's projection which,
although showing the shape of the continents fairly accurately, shows areas with increasing exaggeration as the distance from the equator increases. Great circles on this
projection appear roughly sinusoidal in form as in fig. 5.2.
Other types of projection which are useful in dealing with
great circle radio propagation are discussed below.

An *azimuthal projection* is obtained by projecting the
surface of the globe from a given point on to a tangent
plane. Various different types of azimuthal projection
are obtained depending on where the projection point is
placed. If this point is at the centre of the sphere, a
very useful type of azimuthal projection called a *gnomonic*
projection is obtained. A similar type of map, which is
not strictly an azimuthal projection in the sense in which
we have defined it above is the one which is called an
equidistant azimuthal projection in which places are shown
at the correct distance along the correct bearing from the
centre of the map. Such a map is useful for showing air-
line distances from the central point or, in our case, for
showing the true bearings of different radio stations.
This type of map has the advantage over the azimuthal pro-
jections mentioned above that the whole of the earth's
surface can be depicted on one map even though the distortion of shape and area is very large near the periphery.
On this type of map, of which an example is shown in fig.
5.3, only those great circles which pass through the centre
of the map appear as straight lines. The way in which
other great circles are distorted in this projection can
be seen by studying the lines of longitude (since all longitude lines are great circles).

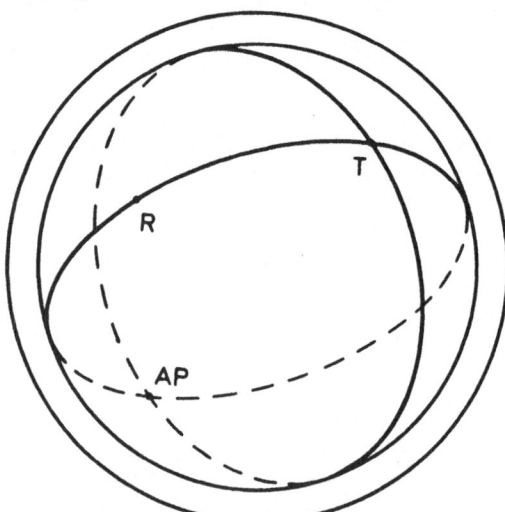

*Fig. 5.1   Idealized earth with concentric ionosphere showing
           great circle paths. AP is the point antipodal to
           the transmitter at T.*

In the above we have found the idealized paths between a transmitter and a receiver. It will be found that a considerable part of the remainder of this work is devoted to a study of how far, in fact, actual conditions deviate from this ideal.

Now we know that the ionosphere is not always favourably disposed towards a radio wave that strikes it. In particular, the regions near the auroral zones are generally disturbed and tend to absorb any radio waves which impinge on them rather than reflect them back to the ground. To illustrate this, we use the Mercator projection in fig. 5.2 showing a series of great circles originating in England. Those regions which are shadowed by the auroral regions are indicated by the dotted lines so that, under conditions of high auroral zone absorption, only in the regions covered by the full lines is reception by direct great circle route possible from a transmitter in England. Note that a large part of New Zealand is in the shadow region.

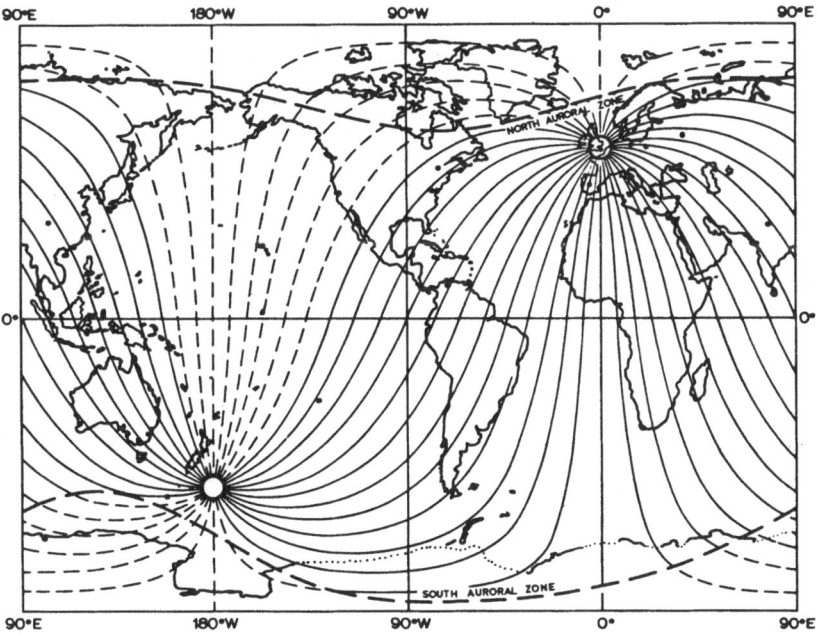

Fig. 5.2 Great circles drawn on Mercator's projection. All of these great circles originate in England. Those regions which are shadowed by the auroral zones are indicated by dashed lines.

It is also generally true that, because the direct radiation from the sun creates the low-level ionospheric layers
(D and E regions) which absorb a considerable amount of energy from any wave passing through them, the night-time
hemisphere of the earth is rather more favourable for long
distance propagation than the sunlit hemisphere.

It is obvious that, were great circle propagation the
only allowed mode, there would be many places between which
communication was usually impossible (depending, among other
things, on the degree of disturbance in the auroral regions).
An important factor which must be considered, therefore, is
the degree to which propagation by paths other than great
circles is possible.

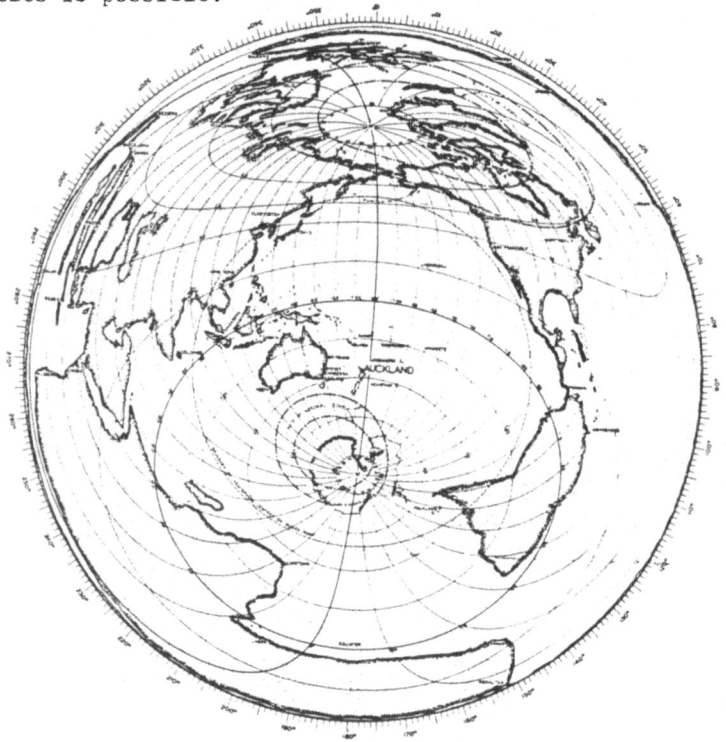

*Fig. 5.3 Equidistant azimuthal projection with Auckland as
the centre. All great circles through Auckland
appear as straight lines. This type of map is most
useful in locating stations from bearing measurements taken at the central point.*

CHAPTER 6

REGULAR DEVIATIONS FROM GREAT CIRCLE PATHS

The fact that, even over a short transmission path, the bearing angle of the received wave can deviate from that which corresponds to the great circle direction has already been mentioned in section 3 in discussing fig. 3.1. It has been found that, when the bearings (each averaged over a period of about an hour to remove the scintillation effects) taken at the same time of the day are averaged over about a month, a fairly regular behaviour emerges. A series of such results obtained during 1953 - 1954 is shown in fig. 6.1. These measurements were made at Auckland using transmissions from two stations located one in Fiji (ZQD) and the other at Brisbane, Australia(VLQ9).

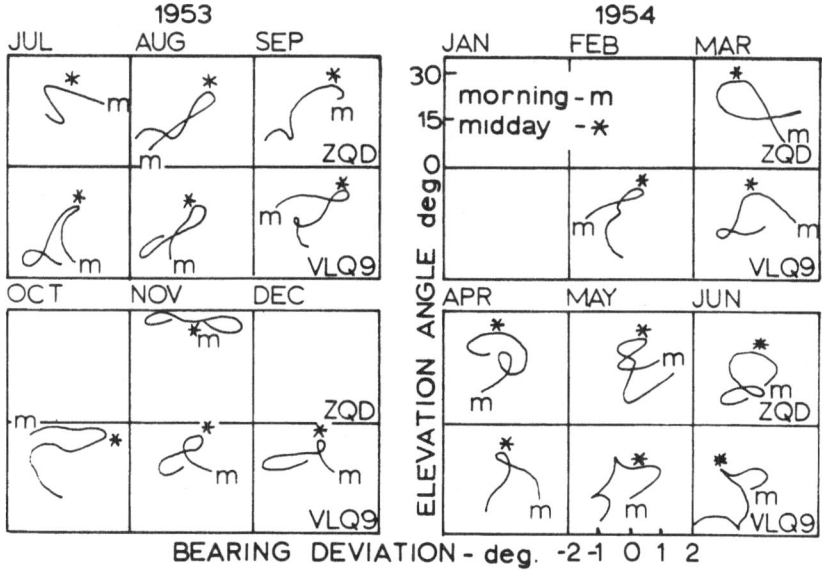

Fig. 6.1   Regular diurnal changes in the direction of arrival of transmissions from ZQD (Fiji) and VLQ9 (Brisbane) observed at Auckland.

Both stations are located about 2000 kms from the
receiving site. It was found that, in both cases, there
was a trend in that the bearing angle tended to increase
at a fairly uniform rate during the day-light hours.
This effect, since it appears in the averaged results,
should be attributable to the predictable large-scale
behaviour of the ionosphere.

An interesting empirical expression for the magnitude
of the regular diurnal change in bearing was obtained by
Titheridge (1958) using results similar to those in fig.3.1.
He found that, for stations in Fiji and Hawaii over the
frequency range 6 - 15 MHz, that the change in apparent
bearing was about $(1/f)$ degrees per hop per hour (where
f is the frequency in MHz).

For the two stations Fiji and Brisbane we have the
interesting feature that the combination of slight differences in frequency and slight differences in distance are
such that the reflection levels in the ionosphere (as
specified by the electron density) should be almost identical. Since the path to ZQD (Fiji) was almost NS and
the path to VLQ9 (Brisbane) almost EW, it would be expected
that ionospheric conditions occurring near the mid-point
of the former path would occur near the mid-point of the
latter at about half an hour or so later.   Since the two
transmission paths are almost at right-angles, effects
which caused an apparent lateral tilt in the ionospheric
reflecting region on one path would be expected to cause
an apparent longitudinal tilt of the region on the other
path.   This was, in fact, found to be true. Elevation
angle measurements were also made and averaged in the same
way as the bearing measurements; it was then demonstrated
directly from the experimental results that the ionosphere
could be regarded as exhibiting a tilt.   A more detailed
analysis of the propagation paths demonstrated that these
apparent regular tilts of the ionosphere arose from
refraction effects in the F1 region.

This type of variation in bearing angle was classified
as a regular diurnal change in section 3. The deviations
from this i.e. the day-to-day fluctuations are essentially
random in character and can thus be treated only as statistical fluctuations.

CHAPTER 7

RANDOM DEVIATIONS FROM GREAT CIRCLE PATHS

It has already been shown experimentally (fig. 3.1) that, for a short path length, there is a significant random variation of the measured bearing from the long term mean bearing which, in the absence of systematic lateral tilts of the ionosphere, is along the great circle path. The problem to be considered is whether the effects of a large number of small random deviations can build up over a long path to give incoming directions which are significantly different from great circle directions.

As a first step, let us take some short and medium distance stations and investigate experimentally the probability distribution of the *deviation* of the bearing from the mean bearing. The mean is obtained for each hour of the day averaged over a thirty-day interval. A few sample distributions of this type are shown in fig. 7.1.

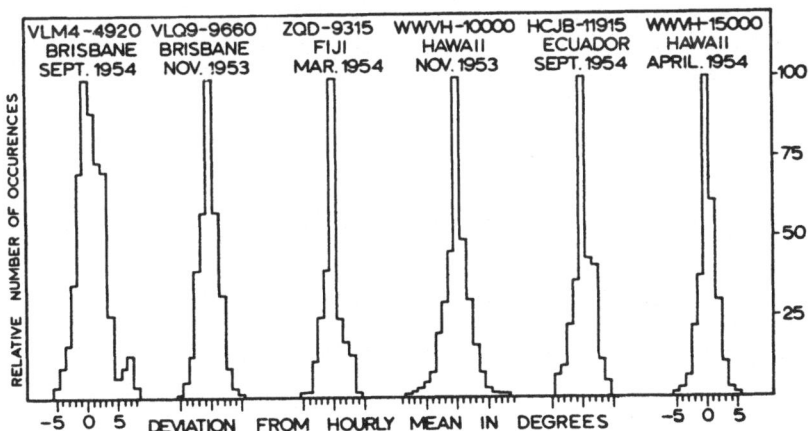

Fig. 7.1 Observed day-to-day variations of the bearing from the hourly mean for the months shown.

These are approximately normal or Gaussian distributions and, as such, can be conveniently described by their standard deviation σ. Write the equation of this probability distribution, which is symmetrical about θ = 0°, as

$$p(\theta) \propto \exp(-N\theta^2) \qquad \ldots\ldots \quad (7.1)$$

i.e. the probability of measuring a bearing in the range θ to θ+δθ is proportional to the right hand side. The parameter N specifies the sharpness of the distribution since the standard deviation is given by

$$\sigma = 1/\sqrt{2N}. \qquad \ldots\ldots \quad (7.2)$$

Note that a small value of N implies a large σ and hence a rather broad flat curve for p(θ) and a large N implies a narrow sharply peaked curve.

Consider a wave which is propagating around the earth in a series of hops between the earth and the ionosphere. Neither the earth nor the ionosphere is a perfectly smooth reflecting surface so that there will be some deviation of the wave from its great circle direction at each reflection point. In this context, we are considering the direction of the wave as a whole i.e. we assume that the scintillation effects have been averaged out. In later chapters, the effects of the *spreading* of the waves at each rough reflection region will be considered further but here we are concerned with the mean direction of energy flow, the mean being taken over a period of half-an-hour or more. This deviation is largely unpredictable in that the exact place of reflection is not known (except perhaps for the first and the last encounter with the ionosphere in a very accurately controlled experiment). Such a controlled experiment could be carried out for example by measuring the direction of the wave reaching the receiver in order to find the place where the signal last encountered the ionosphere. The other part of the experiment would be difficult since it involves specifying the particular part of the transmitted wave which is in the right direction to arrive at the receiver. This means that a very highly directional transmitting antenna system would have to be used. If equipment for measuring directions of arrival is available at both ends of the path, then it is possible, by utilizing reciprocity principles, to transmit from each end of the path in turn and deduce the direction which the transmitted wave followed. It is doubtful whether

the information obtained in such an experiment would justify
setting it up since the path followed by the ray *between*
these end-points could not be obtained in this way. We can,
however, on the basis of the experiments carried out over
short distances, assume with some confidence that the random
variations of direction at a reflection point will be normally distributed. This deduction is possible only because
the short path experiments have shown that the incoming directions are normally distributed about the mean. A simple
geometrical construction indicates that, in such a case, the
variations at the reflection point (assuming a one-hop path)
are also normally distributed. We may note here that, when a
long path is considered so that there are many reflection
points (one at the ionosphere and one at the earth for each
hop) almost any shape of the probability law at each reflection point would yield a resultant directional variation (as
measured at the receiver) which is Gaussian in shape. This
result follows from the central limit theorem which is well-
known in statistical theory. We are concerned mainly with
rather long propagation paths in the course of which a dozen
or more reflections will usually have occurred and we will
make the assumption that each probability distribution of reflection directions is the same (earth and ionosphere reflections included). Even if this is not the case, and there is
certainly a considerable difference between a reflection from
the sea and one from a mountainous region on the earth and
between a reflection from the equatorial ionosphere and one
from the polar ionosphere, we have merely replaced the actual
distributions with an average effective distribution.

We may note that, in some particular cases, specific
areas which are so rough that they introduce special effects
into some propagation paths have been identified from time
to time. In particular, some of the islands off Japan are
thought to be responsible for regular off-great-circle propagation of some signals to Japan.

Assume that the average effective deviation of ray path
at each reflection is described by the equation

$$p(\theta) \propto \exp(- A\theta^2) \qquad \ldots\ldots \ (7.3)$$

where, as in the previous expression, a large value of A in-
dicates a very sharp probability curve i.e. small deviations.
Note that in the previous expression, in which N replaced the
A above, the probability distribution referred to the incoming

bearing directions, not to the deviations at an individual
reflection point as here. The problem in hand, therefore, is
to deduce A from measured values of N or, alternatively, deduce N from assumed values of A.

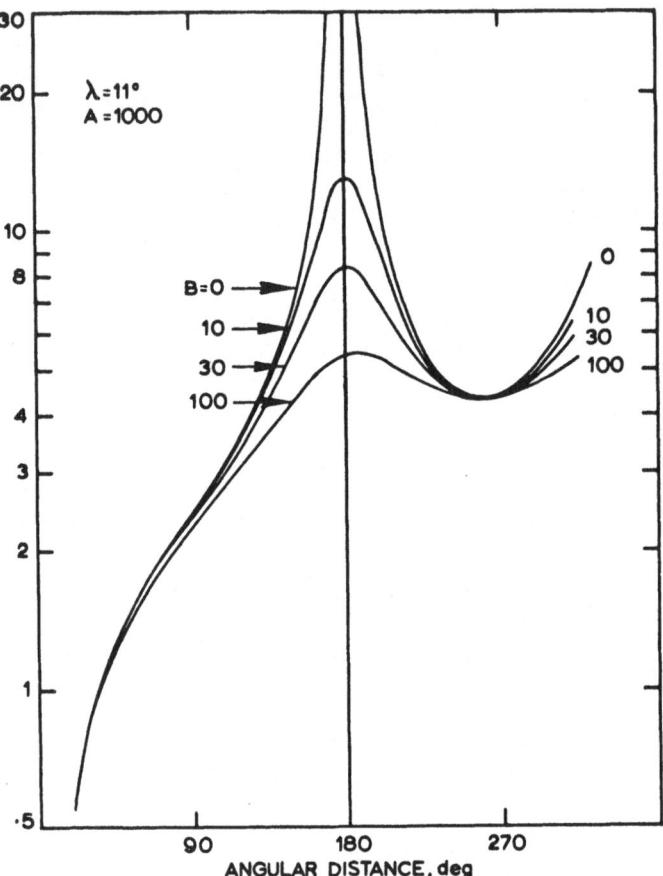

*Fig. 7.2   The theoretical spread of incoming directions as a
function of the angular distance between the transmitter and the receiver.   The parameters B refer to
the beam-width of the transmitting antenna.*
       B = 0 indicates an omnidirectional antenna.
       *B = 10, 30, 100 indicates total beamwidths of
            30°, $17 \cdot 4^o$ and $9 \cdot 5^o$ respectively.*

By considering each reflection point in turn, it is now possible to calculate the probability distribution which would be observed at any distance from a transmitter. There are two methods of tackling the problem. Firstly, we can consider that the standard deviation (or some similar measure) derived from the probability distribution of the deviations produced at each reflection point is a measure of the maximum likely deviation of the ray at that reflection. Then, if all these deviations at successive reflection points are in the same sense (say, the ray always deviates to the left or always to the right) we can calculate the maximum likely deviation in the direction of the incoming ray at the receiver. It turns out, fairly obviously, that the path which such a ray follows is a *small circle* so that the computations are quite simple.

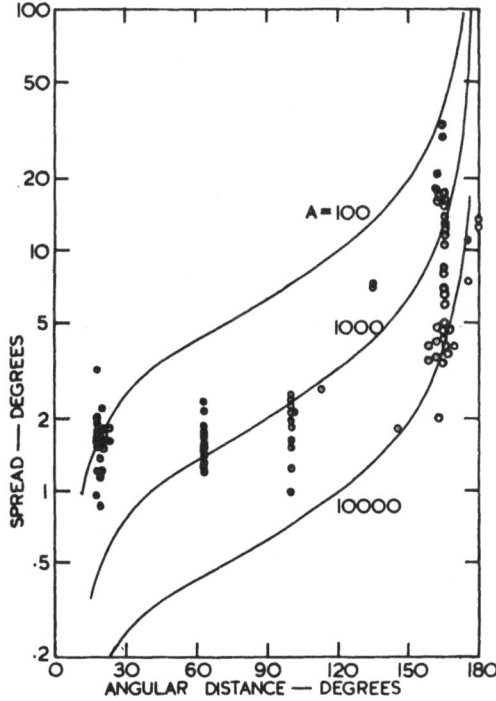

Fig. 7.3 Experimental observations of the day-to-day spread in received bearing angle. The parameter A is related to the spread of the bearing directions at each reflection point.

The second method of approach is rather more complicated mathematically but yields results which do not appear in the former treatment. In this method we calculate, from the assumed probability distribution of the deviations in direction at each reflection point, the actual probability distribution of the received directions. The further important results which are obtained in this treatment provide information on the way in which the beaming of a transmitted signal affects the incoming directions.

We will not consider the first method of treatment here since all the necessary results are included in the second method. The first method is discussed in a paper by WHALE (1959). Neither will we consider the details of the second treatment since it is only the results which we require. The details are published in a further paper (WHALE, 1963). The result which we require at the moment is an expression which relates the probability distribution of the incoming bearing directions to the probability distribution of the deviations at each reflection point (specified by the factor A mentioned above). For an omnidirectional transmitting antenna we obtain the result which is plotted in fig. 7.2 as the curve labelled B = 0. This may be written as:

$$\sigma = 1/\sqrt{2N}$$

where $\qquad N = 2AS_1S_n^2/(nS_1 - S_nC_{n-1})$

$\qquad$ n = total number of steps in the propagation path

$$S_n = \sin(n\lambda)$$

$$C_{n-1} = \cos(\overline{n - 1}\lambda)$$

$\qquad \lambda$ = step length in angular measure.

Notice that a *step* is the length between two successive reflections i.e. either earth to ionosphere or ionosphere to earth. A hop in our nomenclature is thus two steps.

The total angular distance between the transmitter and the receiver is thus $n\lambda$.

The quantity σ can be obtained experimentally as, for example, in the measurements shown in fig. 7.1. By comparing these measurements with the theoretical curves, we

can deduce a value for A, the average effective deviating
factor, which we introduced into the analysis. Since measurements of elevation angle can easily be made with the
rotating interferometer, it is possible to make a good estimate
of the number of hops which have occurred between the
transmitter and the receiver. The curves in figure 7.2 have
been calculated using the reasonably typical value for the
step length of 11°.

A number of experimental determinations of σ are plotted
in figure 7.3 where the theoretical curves are drawn assuming
an omnidirectional transmitting antenna and various values
of the scattering parameter A. It will be remembered that
it requires about a month's results to obtain a sufficient
amount of data to make an adequate estimate of σ and thus to
obtain one point on this figure. Although the experimental
points do not fit any of the theoretical curves very well it
must be remembered that they were obtained at different
seasons of the year and under various different ionospheric
conditions. The indications are that a typical value for
the deviating factor A is given by

A = 1000 (when θ is measured in radians).

This corresponds to a standard deviation of the probability distribution at each reflection point of about 1¼°.

The theoretical curves show that a small amount of
deviation at each reflection is very effective in removing
the absolute control of the antipodal point on the path
followed (mentioned, for example, in section 5) when the
receiver is close to 180° angular distance from the transmitter i.e. close to the antipodal point. It is seen that,
as the receiver is moved further and further away from the
transmitter, the range of bearing angles over which signals
would be expected to be received (at various different times)
increases steadily until, near the antipodal point, they can
be received from any direction at all. The general area
where this effect occurs will be called the *antipodal area*
and is discussed further in Chapter 17. It is important to
note that there are many ways of defining an antipodal area
depending on the particular aspect of its behaviour which
is being considered at the time. At the moment, we are concerned with the day-to-day variations of the direction of an
incoming signal so that we are essentially concerned with a
long-term property of this region. It should be remembered

that the bearing directions with which we are dealing here were obtained by averaging the incoming direction over a period of about half-an-hour to an hour so that the scintillation effects were removed. It will be seen later that these scintillations of the incoming direction may be taken as evidence of the existence of another type of antipodal area i.e. one where actual focussing of the energy from the transmitter occurs.

Here we are considering the deviation of the received signal direction from the *mean* direction which is not necessarily the same as the great circle direction. In other words, the average diurnal variation (averaged over a month) has been removed from the results. This does not affect the calculated value of the spread of angles very much for the short distance transmissions but makes a very great difference at the long distances. If the earth and ionosphere were merely rough although still spherical and concentric and were not absorbing in the auroral and sunlit regions, and if propagation were possible over all paths, then the mean direction would be the same as the great circle direction and the distribution of bearings would be distributed symmetrically about this direction. What we have to deal with, however, is a distribution which is basically of this nature, but out of which large sections have been removed by the presence of the absorbing regions and sometimes by the presence of regions of the ionosphere where the critical frequency is too low to allow propagation to take place.

As the receiver is moved past the antipodal point and receives signals *via* the long path (although this path is now not necessarily a great circle long path) the angular distance increases beyond 180° and it is seen from fig. 7.2 that the range of angles received decreases again. If the experiment were carried far enough so that the receiver was receiving only long-path signals even as it approached the transmitter (from behind, as it were) then it would be expected that a wide range of angles would again be possible. There would thus be a minimum in the range of received angles at some place on the long path between the antipodal point and the transmitter. This appears to occur at an angular distance of about 260° from the transmitter.

We can now investigate what effect, if any, the directivity of a transmitting antenna will have on the range of bearing angles which can be expected. This will also enable

us to predict the general direction from which signals can
be expected to arrive if the transmitter is beamed in some
direction away from the great circle direction to the
receiver.

The analysis is similar to the above. If the transmitter is beamed toward the receiver, the range of bearing
angles to be expected at the receiver is shown by the other
curves in fig. 7.2. We have assumed, as stated previously,
that the effective ionosphere/earth scattering factor A has
the value 1000 and that step lengths are 11° since these
appear to be reasonable average values. A simple single-
lobed antenna beam has been postulated, the beam widths
being specified by the factor B on the various curves. The
actual shape of the main lobe has been assumed to be a
Gaussian curve since this shape agrees reasonably well with
many practical antenna beams (it also, in passing, makes
the mathematics somewhat simpler than would be the case if
some other shape were used). The antenna beam at the transmitter has been assumed to have a power polar diagram which
can be expressed in the form

$$P(\theta) \propto \exp(- B\theta^2) \qquad \ldots \ldots (7.5)$$

where $P(\theta)$ is the power radiated in the small angle between
θ and $\theta+\delta\theta$. We see that this distribution of power is of
the same form as the probability distributions which we have
been considering previously. Typical values of B would be
10, 30, 100 corresponding to total beam-widths in the sense
illustrated in fig. 7.4 of 30°, 17.4° and 9.5° respectively.
The latter case would represent an unusually directive antenna for a station normally engaged in short-wave broadcasting. Beam-widths of 20-30° (corresponding to B = 10-30)
are much more common.

It is necessary to enquire at this stage whether it is,
in fact, permissible to equate the probability distribution
of directions with a distribution of power, as we have done
in order to obtain the results illustrated in fig. 7.2. We
have, in fact, regarded equations (7.3) and (7.5) as both
describing the probability that a ray will travel in a particular direction θ. We note that the probability distribution of incoming directions is essentially a graph of the
relative time that an incoming bearing would be found in the
small range of angles θ to $\theta+\delta\theta$ and the magnitude of this
relative time is linearly proportional to the probability

p(θ). This relative time of occurrence need not occur in one unbroken period; it will, except very near the tails of the distribution, be made up of a large number of small intervals which are simply added together. The fact that the individual contributions of time add linearly indicates that the probability distribution behaves in the same way as a power distribution (where individual contributions of power add linearly) so that the two distributions may be regarded as equivalent.

It can be seen from the curves in fig. 7.2 that the effect of the transmitter beam-width is relatively unimportant except very close to the antipodal point. It is interesting to note that, near the 260° position, the transmitter polar diagram has no effect whatsoever on the spread of directions. This particular position is discussed in more detail below.

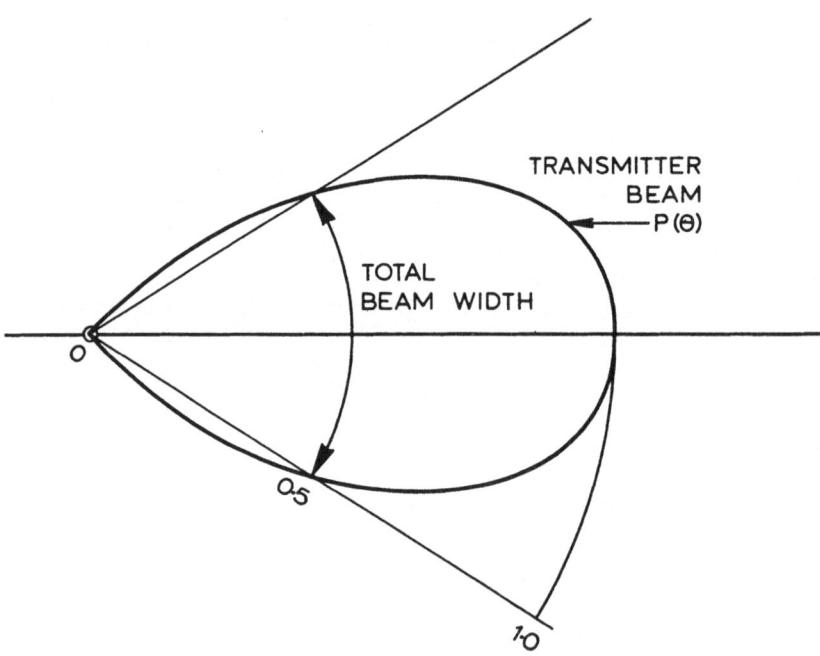

Fig. 7.4 Specification of the total beam-width of an antenna as the angle between half-power points.

*Fig. 7.5 Relation between the beaming direction (θ_T) and the
angle of received maximum likelihood of signal
direction (ξ_{max}).*

A further result which can be deduced from the theory
is the way in which the transmitter beaming *direction* affects the average direction from which signals may be expected. Again, the results are best presented in the form
of a diagram and again they are based on A = 1000 and
$\lambda = 11°$. In fig. 7.5 the line TR is the great circle between
the transmitter and receiver. The transmitting antenna is
beamed so that its maximum is in a direction making an angle
θ_T measured from the great circle. Then the probability distribution of the range of bearing angles which would be expected at the receiver still has the shape which is defined
by the standard deviations given in fig. 7.2, but the maximum of this probability distribution is offset from the
great circle direction by the angle ξ_{max}. The ratio of
these two angles, i.e. ξ_{max}/θ_T, is plotted in figure 7.6
for various beam-widths of the transmitting antenna.

If the receiver is at the point antipodal to the transmitter there is always a one-to-one relationship between the
transmitter beaming direction and the maximum of the received
directions, since there is no preferred great circle path.
At other positions the curves in fig. 7.6 may be thought of
as indicating the degree to which the great circle controls
the path i.e. the "tautness" of the path. The path is taut
when ξ_{max}/θ_T is small, i.e. there is little deviation of the
maximum of the probability distribution of the incoming directions from the great circle direction. It will be noticed
that this is generally the case. That is, for the usual
beam-width (B = 10-30) associated with transmitting antennas

in the range of frequencies which we are considering (5-25 MHz), the beaming direction will have a relatively minor effect on the received bearing.

Again, we find that there are particular properties associated with the 260° position. Referring to figs. 7.6 and 7.2 it is seen that, at this distance from the transmitter, the incoming waves are spread about a maximum which is always in the direction of the great circle bearing to the transmitter irrespective of the beaming direction of the transmitter and also, as mentioned previously, that the spread of angles does not depend on the width of the transmitter beam.

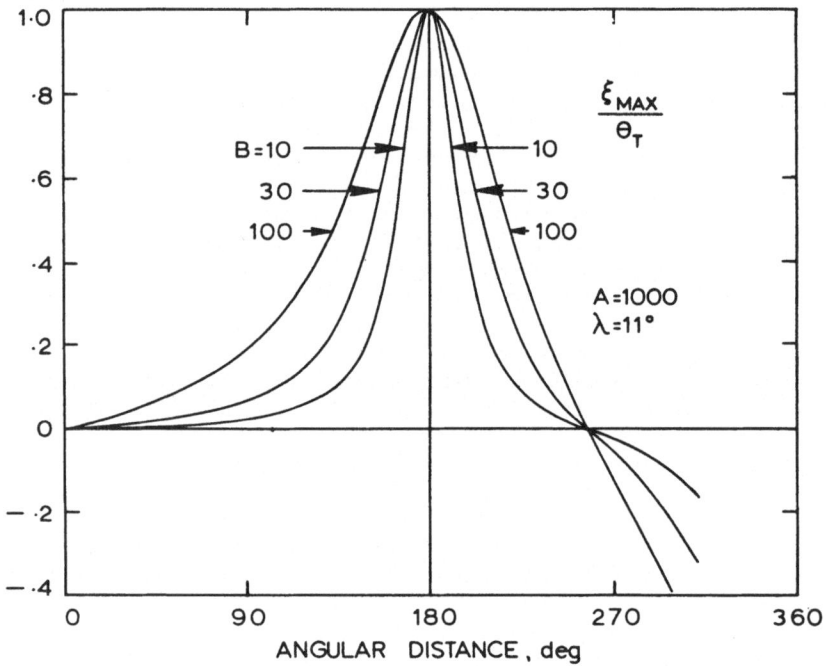

Fig. 7.6   The theoretical shift of incoming bearing angle with changing beaming direction of the transmitting antenna.   The parameters B refer to the beamwidth of the transmitting antenna.

As an indication of the way in which the deviating fac-
tor A affects the "tautness" of the path, the ratio ξ_{max}/θ_T
is plotted in fig. 7.7 as a function of A. In this diagram,
we have again taken $\lambda = 11^\circ$ and have assumed a value of
about 17° (B = 30) for the total beam-width of the transmitting antenna. We see that with large values of A (i.e.
small deviations at each reflection point) the incoming
direction is close to the great circle direction except for
those cases where the receiving station is nearly antipodal
to the transmitter. On the other hand, for small values of
A (say A = 100) the received direction can be markedly affected by the transmitter beaming direction. We would thus
regard the first case as a "taut" path while the latter
would not be in this class.

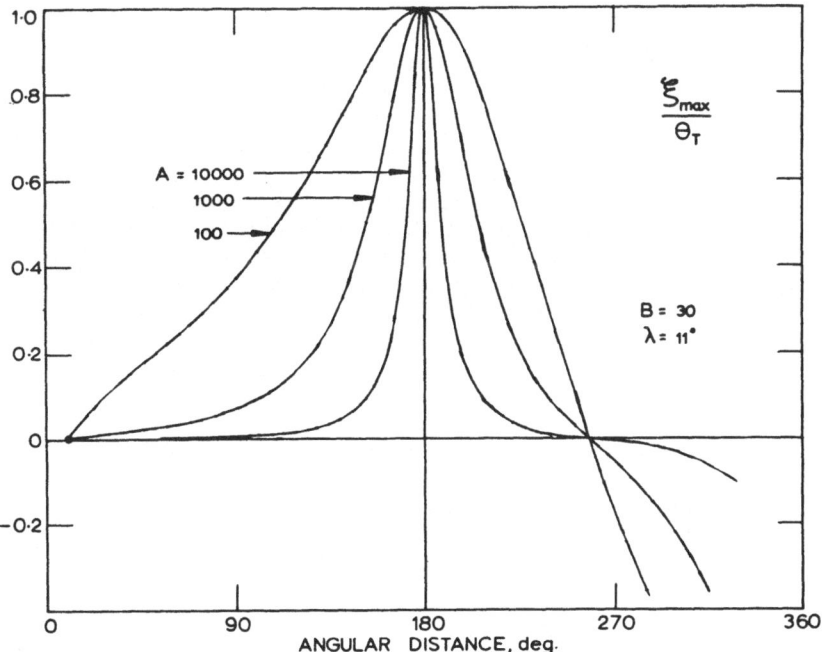

*Fig. 7.7 The way in which the ratio (ξ_{max}/θ_T) changes with
the scattering parameter A.*

It must be emphasized at this point that, in all the
above, the rather idealistic assumption has been made that
all great circle paths are possible. Over relatively short
distances, say up to 90° angular distance, experimental conditions can be chosen so that this assumption is not serious-
ly violated. Under these circumstances the ratio ξ_{max}/θ_T is
small and the deviation of the average direction from the
great circle is relatively unimportant. At receiving points
approaching 180° angular distance from the transmitter where
the deviation becomes appreciable the fact that some great
circle paths can not be sustained because of absorbing or
other mechanisms may assume overwhelming importance.

It is concluded that, in most cases of practical importance, the direction of beaming of the transmitting antenna
will have only a small effect on the direction of the received waves; it may, however, have a very great effect on
the *strength* of the received signal.

The experimental results in connection with this problem are to some extent inconclusive but are nevertheless interesting in that they indicate the degree to which practical
conditions depart from the idealized conditions which have
necessarily been assumed in obtaining the above theoretical
results. While a change in the transmitter beaming direction
will have little effect other than to change the received
signal strength in the case of a short path transmission, it
is possible that, over the longer paths, such a change could
give rise to measurable changes in the direction of the incoming signal. The results of such an experiment will depend on how many propagation paths are possible between the
transmitter and the receiver. If there is only one possible
path then a change in the transmitter beaming direction can
only alter the received signal strength but, if several
different paths are simultaneously possible, a change in the
beaming direction may cause a change in the received direction. There is, of course, some practical difficulty in deciding what paths are, in fact, possible.

Some typical results of an extensive investigation into
the effect of the transmitter beaming direction on the bearing of the received signal are plotted in Fig. 7.8 which is
representative of the measured values of the bearing angle of
BBC short-wave braodcasts on frequencies between 9 MHz and
21 MHz. Each circle represents the received bearing averaged
over about 1 hour, the size of the circle being proportional

to the number of times a particular bearing was observed
during the period of the investigation. All these measurements were made in 1956-1957. The plot shows the departure of the received bearing from the great circle bearing
(taken as 170°/350°E), without regard to the sign of the
departure so that the scale vertically is 0-90°, as a function of the difference between the nominal transmitter
beaming direction and the great circle bearing of the receiver from the transmitter (taken as 190/10°E), again
without regard to the sign of the difference, so that the
horizontal scale is also 0-90°. The diagonal dashed line is
the locus of bearings which are the complement of the transmitter beaming directions; in the idealized case of a transmitter situated *antipodally* to the receiver on a smooth
spherical earth surrounded by a smooth spherical ionosphere
concentric with the earth, this line would give the received
bearing for any transmitter beaming direction. The horizontal line forming the base of the diagram is the locus of
bearings which lie along the great circle direction. In
general, it has been found that most of the points lie in
the bottom right-hand half of the diagram, i.e. most of the
signals arrive from some direction between the great circle
direction and the complement of the transmitter beaming
direction.

If the deviations from the great circle direction shown
in fig. 7.8 do, in fact, arise from transmitter beaming effects, then the value of the radio ξ_{max}/θ_T can be estimated
from the slope of the best straight line which passes through
the origin and through the bulk of the experimental points.
Such a line would have a slope of about ·3 - ·5. The expected value would be given by fig. 7.6 at the angular distances of 164° and 196° (the short and long path distances
between the transmitter and the receiver) and is certainly
of this order. We have made no distinction between short-
path and long-path propagation in this investigation since
the curves in fig. 7.6 are asymmetrical about the 180° line
for angular distances close to this. It should be realised
that the above type of experiment does not necessarily prove
that the bearing direction is dependent on the transmitter
beaming direction. It really only shows that the bearing
shows a tendency towards this great circle bearing.

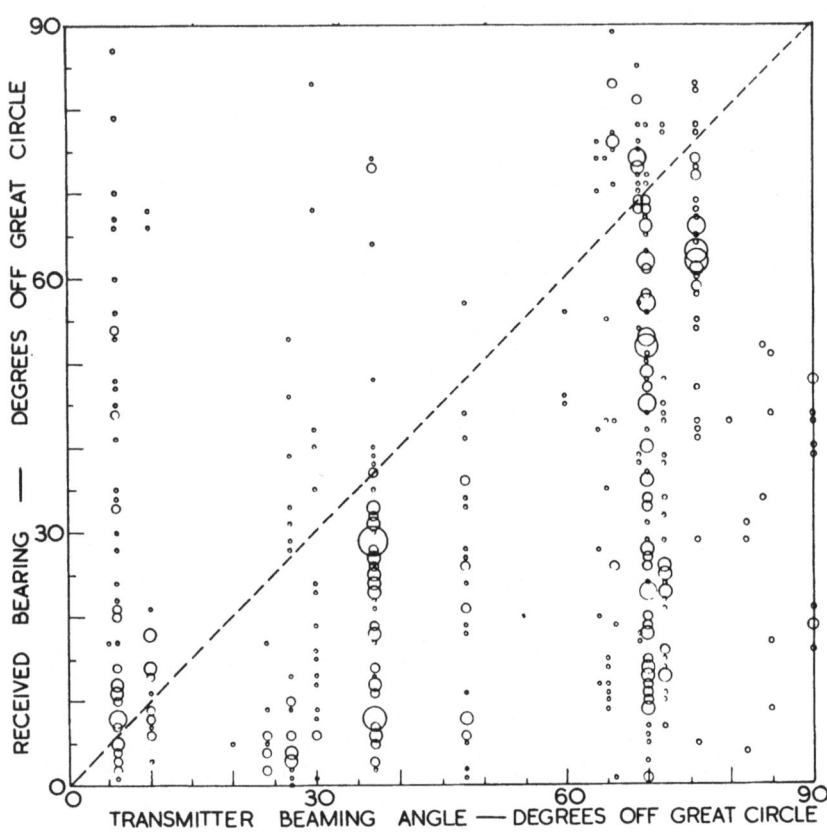

Fig. 7.8 Relation between the experimental received bear-
          ing direction and the transmitter beaming direc-
          tion. Both quantities are measured using the
          great circle direction as zero. The size of the
          circles is proportional to the number of
          occurrences.

A more significant series of results was made possible
when the British Post Office arranged for the transmission
from Slough of a fixed frequency which was radiated first
from one antenna and then from another during seven days between 24 October and 11 November, 1955. In these experiments the two antennas employed were firstly an almost omni-
directional antenna (horizontal dipole) and secondly a
rhombic antenna beamed in the direction 310°E. This is 60°
off the great-circle direction and should thus be effective
in demonstrating any change of the received direction with
change of the transmitting direction. The average over the
7 days of the value of the quantity {(bearing for rhombic
transmitting antenna) - (bearing for dipole transmitting
antenna)} is plotted as a function of time in fig. 7.9 for
receiving stations both at Seagrove (Auckland) and at Awarua
(Invercargill). This latter station is at the very southern
end of New Zealand and is much closer to the point antipodal
to the transmitter than is Auckland. Awarua is 9° angular
distance from the antipodal point while Auckland is 15½°
distant.

The measurements at the Awarua station were made by
personnel of the New Zealand Post Office using an Adcock
type direction-finder. There is a small effect on the re-

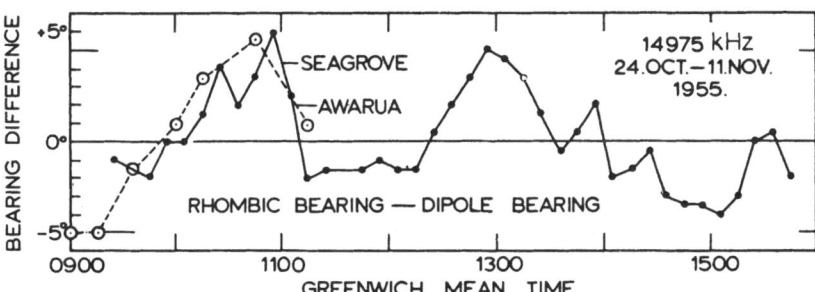

Fig. 7.9 The observed difference between the bearing angles
          of signals received when the transmitting antenna
          is changed periodically from a directional (rhombic)
          system to a relatively non-directional (dipole)
          system. The results on the same transmissions meas-
          ured at two different receiving sites are shown.

ceived bearing (consistent between the two stations) arising from the change in antenna directivity but not consistent in magnitude or direction. As this change is only of the order of a few degrees it would be insignificant when considered in relation to the usual large variations of these bearings.

Some further results obtained on BPO transmissions, in which the bearing at the time of changeover of the transmitting antennas was observed very closely, also show negligible change of the received bearing, even when the transmitter beaming direction was changed by 180°. A sample record of the bearing obtained during such a changeover is reproduced in fig. 7.10. In this case the receiving site was at Seagrove, Auckland. It was found, however, that at Awarua there were some occasions when a very large change of bearing occurred as the transmitter beaming direction was altered.

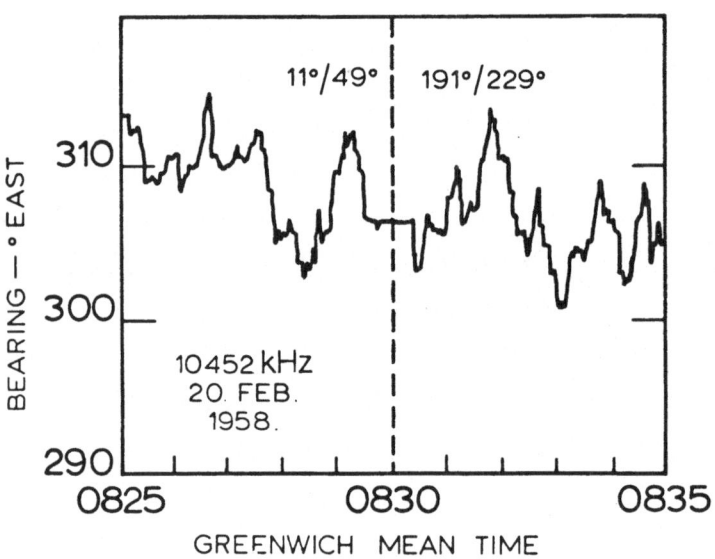

Fig. 7.10 Detail of the bearing fluctuations at Auckland as the beaming direction of the transmitting antenna was changed.

No equivalent experiments at shorter distances and free of the complicating effects arising from the fact that the transmission path passed through or close to the auroral regions appear to have been carried out.

In the light of the above results which indicate that in practice, at least in the case of propagation paths which lie close to the auroral regions, the transmitter beaming direction has a relatively minor effect on the received bearing angle, it is now relevant to enquire into the likely effect of such an absorbing region on the received bearing.

CHAPTER 8

PRESENCE OF AN ABSORBING REGION

The general theory which has been presented above takes no account of highly absorbing regions.

Such regions will be present, for example, near the polar regions where there is generally a ring of high absorption roughly at the same place as where the maximum number of aurorae occur, the auroral zone. Again, in the polar regions it is common for the ionospheric absorption to be considerably above the world-wide average at times of high sunspot number in regions centred approximately on the geomagnetic poles. This is usually referred to as polar cap absorption and occurs at geomagnetic latitudes greater than about 65°.

From an observational point of view, the detailed process whereby the energy is lost is immaterial so that a "thin" patch of ionosphere which allows the radio wave to penetrate it and thus become lost to space can also be regarded as an "absorbing" region. Such a patch could occur where the critical frequency of the ionosphere were abnormally low.

There is always some loss of energy when a wave is reflected from the ionosphere. The general effects of this normal loss can usually be allowed for and, in practice, the absorption sets a lower limit to the frequency which can be employed on any propagation path. Those places where the absorption is so high that a negligible amount of radio energy passes through, may be regarded as behaving like opaque *stops*, to use the analogous optical term.

We can then use the previous theory to investigate the distribution of received rays when a highly absorbing region is inserted between the transmitter and the receiver. Consider the absorbing region to be the perfectly opaque "stop" represented by the heavy curved line with its edge at E in fig. 8.1. We will consider two cases viz. that shown in fig. 8.1a, where the stop is near to the region antipodal to the transmitter (AP is the point which is the antipode to T) and that in fig. 8.1b, where the stop is in the same hemisphere as the transmitter. The transmitter is at T and the shadow

regions (for great circle propagation) are shown as shaded
areas. These two cases correspond to the two most commonly
occurring practical problems of absorption by the two auroral
zones. For the specific case of transmission from Europe to
the Pacific area, there will usually be a large number of
important paths which pass through one or the other of these
zones. Note that, because the auroral zones are situated
almost symmetrically around the geomagnetic poles which are
themselves almost antipodal points, if a short path great
circle route passes through one of the auroral regions, then
the long path route will pass through the other. These are
the two cases which have been idealized in the present discussion.

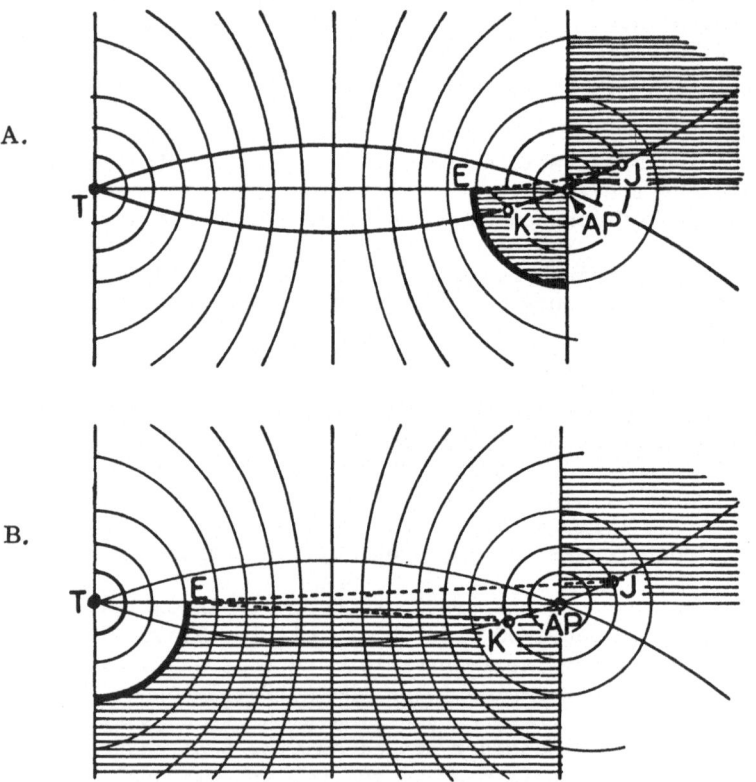

Fig. 8.1 Shadowing effect of an absorbing region placed
          near E. The transmitter is at T with its anti-
          podal point at AP in both cases.

The spread of received angles arriving at E in fig. 8.1 can be found from fig. 7.2. Then, the spread of angles leaving E is this incident spread modified by the extra deviations introduced by the reflection at E and specified by the factor A pertinent to the position E. For places within the shadow region, then, we can regard the transmitter at T as being replaced by a virtual transmitter at E with a polar diagram (interpreted as the probability of emitting a ray in a particular direction) specified by the same expression as has been used previously.

In the situation shown in fig. 8.1a the receiver will usually be not far distant from the absorbing stop i.e. the angular distance between the virtual transmitter at E and the receiving point will be small so that, using fig. 7.6, it is seen that the direction of the received signal will not vary very much from the directions of the edge E of the stop. In making this deduction, we have noted, from fig. 7.6, that the direction of the maximum of the probability distribution is not displaced very far from the great circle direction to the transmitter (in this case the virtual transmitter) when the angular distance from the transmitter to the receiver is less than about 90°.

In case b the distance from E to the receiver may be much greater so that, in this case, the received direction may differ from the great circle direction to the virtual transmitter by a larger amount. This conclusion again follows from the results drawn in fig. 7.6 since, firstly, the displacement of the maximum is predicted to be relatively large at angular distances approaching 180° and secondly, the probability distribution of directions at the virtual transmitter position may be quite narrow (from fig. 7.2), since the stop edge is close to the actual transmitter at T. In this case, the appropriate curve in fig. 7.6 is one corresponding to a narrow-beamed virtual transmitter and it is seen that the deviation increases as the transmitter beam-width decreases. On the other hand, the auroral regions or any other absorbing region may be relatively disturbed so that the deviations introduced at this particular reflection point could be considerably greater than the average which has been used in the analysis. It would appear, therefore, that the received direction, even in this case, should approximate to the direction of the edge of the stop.

This type of behaviour, where the measured directions
indicate the direction of the edge of the absorbing stop is
in many ways equivalent to the phenomena observed in the
simple physical experiment shown in fig. 8.2. Here we have
a light source at S illuminating a ground glass screen EG
with an opaque stop below E. The scattering by the ground
glass causes the transmitted light to travel in a range of
directions which is clustered about the incident direction.
This corresponds to the probability distribution of the
angles through which the radio wave may be deviated. We
may easily find the direction of the incoming light at the
points a, b and c by placing an object there and observing
the direction of its shadow. The shadows appear as shown in
the diagram. Although these are always indistinct because
the light source (the illuminated ground glass screen) is
very broad, they show the general characteristics which we
have deduced above. That is, if the object is in the "illuminated" region (a and b) the shadow direction indicates the
direction of the source at S. If the object is in the
"shadow" region (c), its shadow indicates the direction of
the edge of the stop at E. It must be remembered that,
although this experiment is certainly relevant, the distributions with which we are dealing in the radio case are
obtained on a long term basis i.e. they do not correspond to
any real distribution of power or light intensity which is
observed here. To anticipate some later work, we may say
that, if the ground glass screen were changing its detailed
structure with time, we would be observing fine scale scintillation effects in the above experiment. One of the major
differences between these effects and the directional effects
we are considering is that, in the case of scintillations
there is usually an undeviated wave (the specular component)
to be considered whereas the long term directional deviations
considered at the moment have been found to have an approximately gaussian distribution i.e. there is no undeviated
component.

It is found, in the above experiment, that the shadows
cast by the object are somewhat asymmetrical in shape.
This is confirmed by calculated probability distributions of
incoming directions for a fixed position of the stop edge.
It is known that the auroral absorbing regions change position throughout the day and expand and contract from day to
day. Since the probability distributions with which we are
concerned are obtained by analysing the day-to-day variations
of incoming directions from the mean direction we should not

expect to find this asymmetry in the distributions obtained experimentally. In fact, the distributions are so widely spread that it is difficult to specify their statistical parameters accurately.

We may summarize the above with the statement that, in the case of near-antipodal propagation, if conditions over most of the propagation path are such that the preferred path is through the auroral regions, the received direction will be close to the great circle direction of an edge of the auroral absorbing region as seen from the receiver.

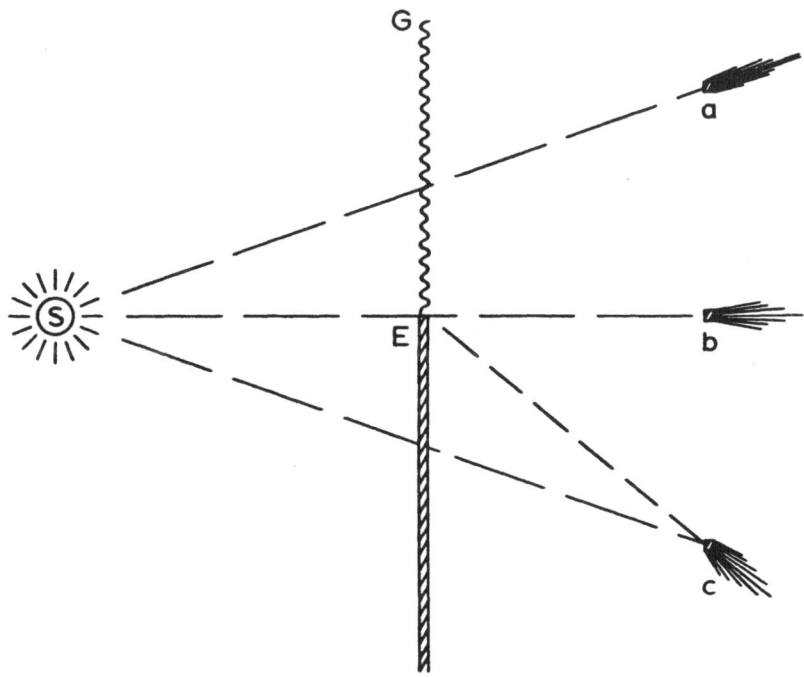

Fig. 8.2   A simple optical analogue of scattered rays being "bent" around an opaque screen. The shadows cast by the objects at a,b,c are indications of the direction of travel of the rays from the ground glass screen GE illuminated by a source at S.

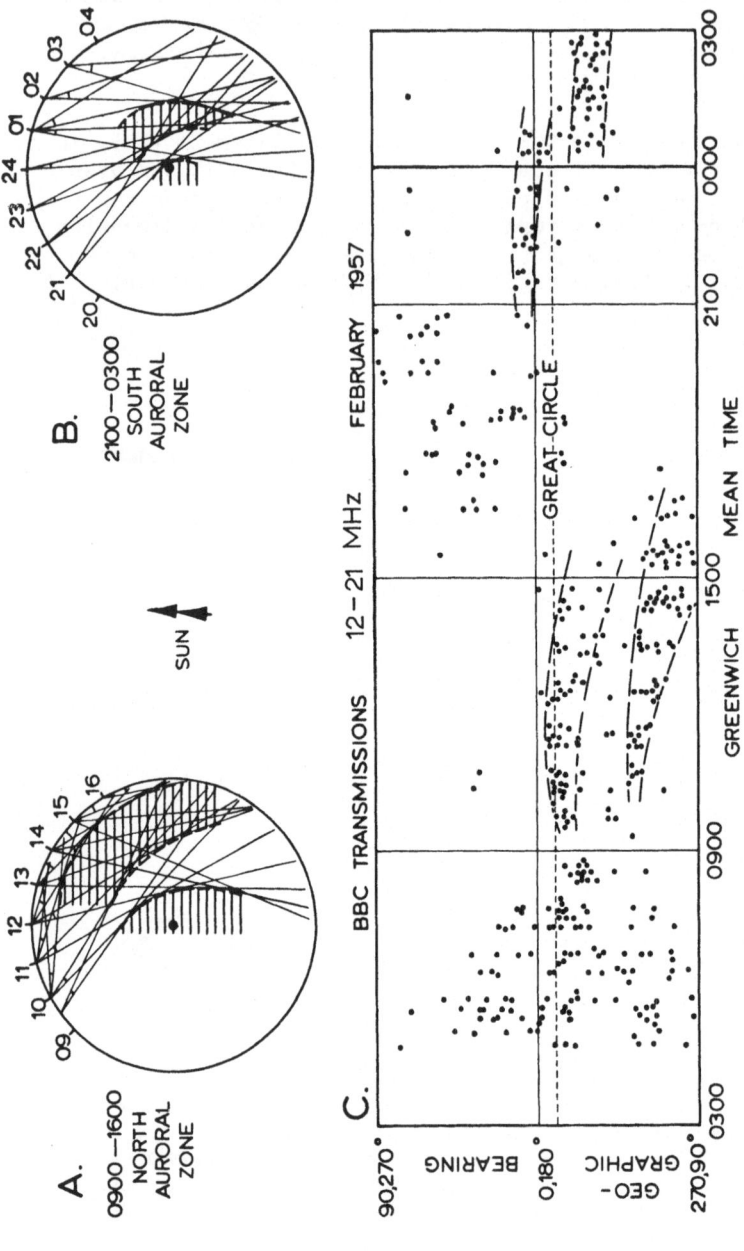

Fig. 8.3 Auroral zone absorbing regions deduced from bearing measurements of a distant transmitter.

Now the auroral regions, although varying in extent, are limited, in general, to latitudes greater than about 50°. Furthermore, on any particular day, the usual variation of the size and position of the auroral absorbing region should not be extremely violent. With these limitations in mind we can investigate the experimental directional observations with a view to specifying those times when the directions are controlled by the presence of the auroral regions.

We have already deduced from a study of the day-to-day variations in the direction of signals propagated over short and medium paths (under conditions where the auroral regions do not affect the propagation) that a very large spread of the received angles is to be expected at near-antipodal distances. Since the effect of the presence of any absorbing region will be to reduce this spread we may expect that there may be some periods of the day when these effects of the auroral regions can be identified. We will take, as an example of the observed results, those shown in fig. 8.3C. These are the received bearings (averaged over about one hour) of BBC stations operating in the frequency range 12-21 MHz during February, 1957 and measured at Auckland. The receiving site is about 15° angular distance from the antipodal point of the transmitters. Since a number of different transmitters located at different places in England were utilized, all the paths involved are not identical but any effects arising from this fact are almost certainly negligible.

There are two periods during the day when the received bearings are restricted to a moderate range and change in a reasonably regular fashion. These occur between 0900 and 1500 G.M.T. and between 2100 and 0300 G.M.T. It is possible that, at these times, the auroral regions are the controlling factors. Further consideration of these two periods is given in the next chapter.

The other two periods i.e. from 0300 to 0900 and from 1500 to 2100 G.M.T. are times when the received bearings vary widely. Because of the large scatter in the results, it seems doubtful whether any method of predicting incoming bearing directions during these periods could be devised.

CHAPTER 9

LOCATION OF THE ABSORBING PARTS OF THE AURORAL REGIONS

The average auroral zones, as found from measurements
on the visible aurora, are nearly circular bands approximately centred on the geomagnetic axis poles. The position
of the line of maximum aurora, when measured at a fixed
location, appears to move N and S during a 24 hour period,
being furthermost from the equator at midday. From these
well-known observations, it seems probable that the
auroral region, of some relatively stable shape, pivots
around the geomagnetic pole retaining a fixed aspect of
the sun.

In order that the shape of the auroral zone may be
found from the measured bearings, these must first be
converted into actual ray-paths near the magnetic poles.
A suitable projection for this is that polar gnomonic
projection which is tangential to the earth's surface at
the *geomagnetic* poles. This type of projection has the
property that all great circles become straight lines and
hence the plotting of ray-paths is greatly simplified.
To find the shape of the absorbing zone we must assume that
it remains constant for the period of the observations and
that it remains fixed in position relative to the sun with
the receiving point thus apparently moving round it once
in 24 hours.

During the daytime at the receiver, most of the rays
from European stations received in Auckland pass close to
the South auroral zone while during the night most of them
pass close to the North Auroral zone. All great circles
through the receiving point pass through its antipode so
that the pattern may equally well be referred to this antipodal point, the bearings of rays arriving from near to the
North auroral zone being plotted on the same projection as
those from the South. A 12 hour adjustment in the time is
necessary, however, if the orientation to the sun is to be
preserved. It has been suggested that the auroral effects
arising from positive particles may be different from and
occur at different places from those arising from negative
particles so that it is possible that the auroral zone is
asymmetrical about a line joining the sun to the earth.
However, the time adjustment of 12 hours ensures that the

orientation of the pattern is preserved, the plotted shape
being a satellite's view of the Southern auroral zone and
a seal's view of the Northern auroral zone.

All the measured bearings of BBC short-wave broadcasts
received at Auckland during February 1957 are plotted in
Fig. 8.3C. The diagram has been compressed vertically,
bearings and their reciprocals being plotted on the same
vertical scale; the signals in the period 0900-1500 GMT
arrived from the north while those in the period 2100-0300
GMT arrived from the south. The dashed lines which have
been drawn to enclose the most common bearings found during
these two periods have been used to construct Figs. 8.3A
and 8.3B. The measured bearings have been converted to
geomagnetic bearings, taking the geomagnetic poles as being
at $(78.5^{\circ}N, 78^{\circ}W)$ and $(78.5^{\circ}S, 102^{\circ}E)$, and plotted on the
gnomonic projection. The radius of the boundary circle
represents the distance of the receiving point or its
antipode from the nearer geomagnetic pole and is 50.3° of
angular distance in each case.

From previous measurements made by workers in Canada
and elsewhere it seems that the auroral absorbing regions
may occur as relatively narrow bands. The signals could
thus cross them at right angles fairly easily if the ionosphere reflection points occurred on each side of the
absorbing band so that the wave effectively straddled the
zone, pointing to the shaded portion in Figs. 8.3A and
8.3B as the measured absorbing regions. It is immediately
obvious that, while of similar shape, the two diagrams are
very different in size indicating that during this period
the North and South auroral zones were of different size or
were not similarly situated with respect to the geomagnetic
poles. This investigation refers to the average shape of
the auroral zone during the month of February 1957.   The
variation of shape, size and location with the season,
magnetic activity, etc., requires further investigation.
Figure 8.3C is typical of the other months for which a
sufficient number of bearings to produce a pattern has been
obtained in that the distributions for the two most clearly
defined periods, viz. 0900-1500 GMT and 2100-0300 GMT, are
often of different shapes.

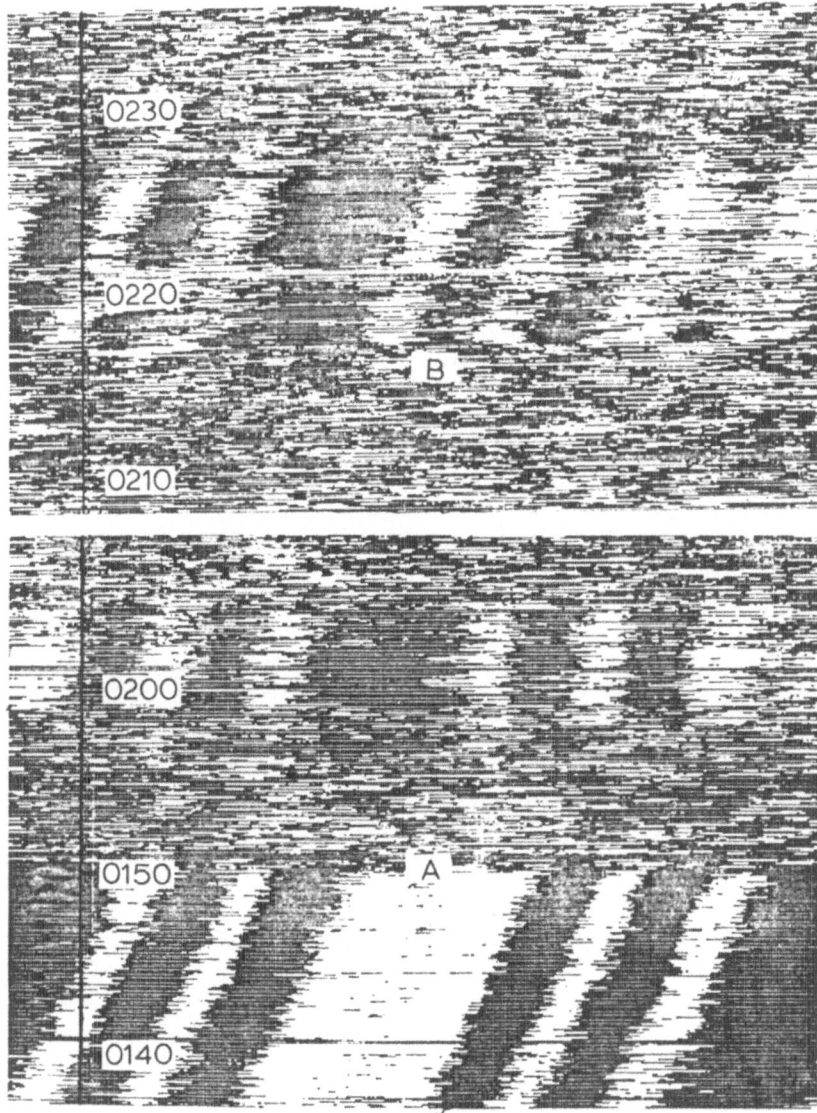

Fig. 9.1 Record obtained on the fast interferometer of the signals from the Russian satellite VOSTOK II. The great circle transmission path entered the Southern auroral zone at A and left it at B.

A particularly interesting demonstration of the way in which the Southern auroral zone obstructs the passage of radio waves was obtained by Dr J.E. Titheridge during the transmissions on about 19.994 MHz from the Russian spacecraft VOSTOK II. The satellite's orbit was such that it circled New Zealand and periodically passed behind the auroral zone. One such passage is shown in fig. 9.1; the record was taken on the fast interferometer where the satellite's trace disappeared at A as the great circle transmission path entered the auroral region and reappeared at B as the transmission path left the vicinity of the auroral region. This behaviour occurred on each occasion that the satellite passed behind the auroral zone, the sharpness of some of the transitions (e.g. the one shown at A) being quite dramatic.

CHAPTER 10

SEASONAL VARIATIONS IN DIRECTION OF ARRIVAL

It is obvious from fig. 8.3 that the scatter of the received bearing directions is very large, especially during the time periods 0300 to 0900 and 1500 to 2100 GMT. Nevertheless, it is possible that some limits to this variation could be indicated by an investigation of the average behaviour over a long period. Such a series of results is presented in figs. 10.1, 10.2, 10.3 and 10.4. These figures show the relative number of times in particular bearing was received during the years 1953-1957. As many measurements as possible at any frequency within the range 12-21 MHz were made. The main factor limiting the number of results obtained was the cost of such a concentrated programme. The four sets of results correspond to the same four time-periods that appeared to show the different types of behaviour indicated in fig. 8.3. All the results shown in figs. 10.1 - 10.4 have been gathered together and included in the plots of bearing angle *vs* month of the year shown in figs. 10.5 - 10.8.

While there are some trends observable in these results, it is obvious that considerably more data is required before any definite conclusions as to the average seasonal behaviour can be drawn. There was, however, one set of results which showed a well-defined seasonal variation of direction. These bearings (fig. 10.9) were all made at about the same time (near 2100 GMT) on the same transmission. The mean direction is shown in the lower portion of the figure with a sine wave of $23\frac{1}{2}°$ amplitude for comparison. In this particular case it appears that the propagation path was controlled mainly by some type of twilight zone effect (see DAVIES, 1965).

It can be concluded, from the above, that it is most difficult to predict the path which a wave will follow when the transmitter and receiver are nearly antipodal and that it seems that a considerable amount of further work in this field is needed.

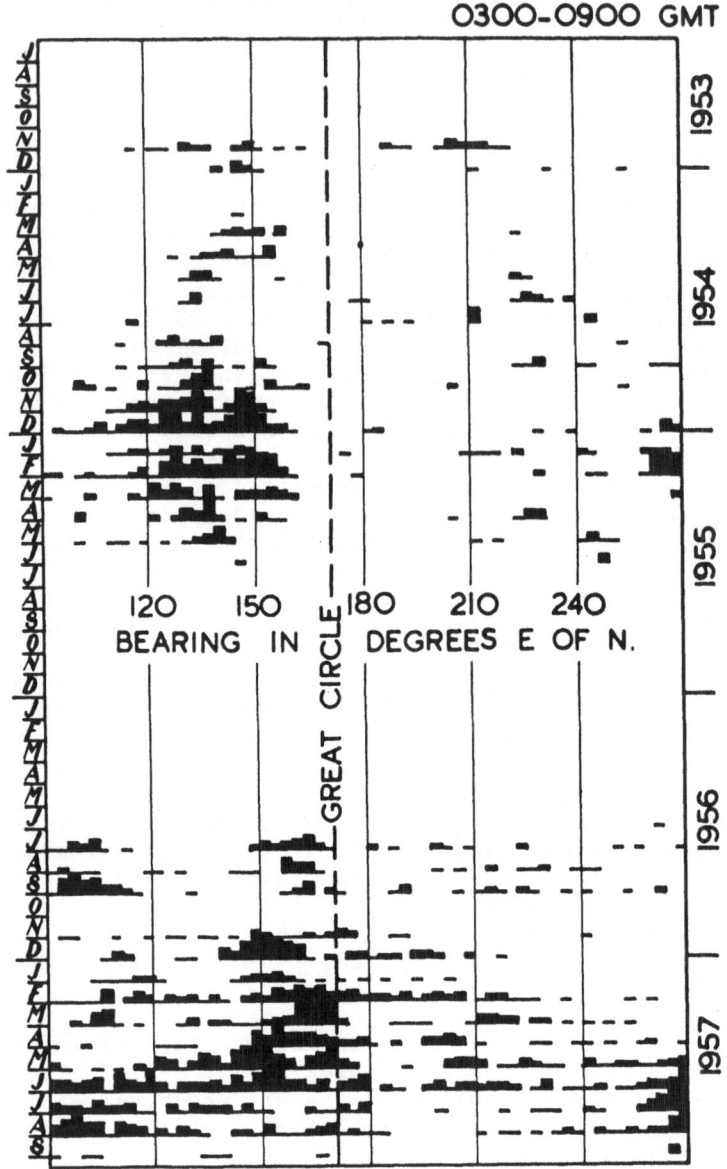

*Fig. 10.1   Observed bearing angles over the years 1953-1957
during the 6-hour period 0300-0900 GMT.*

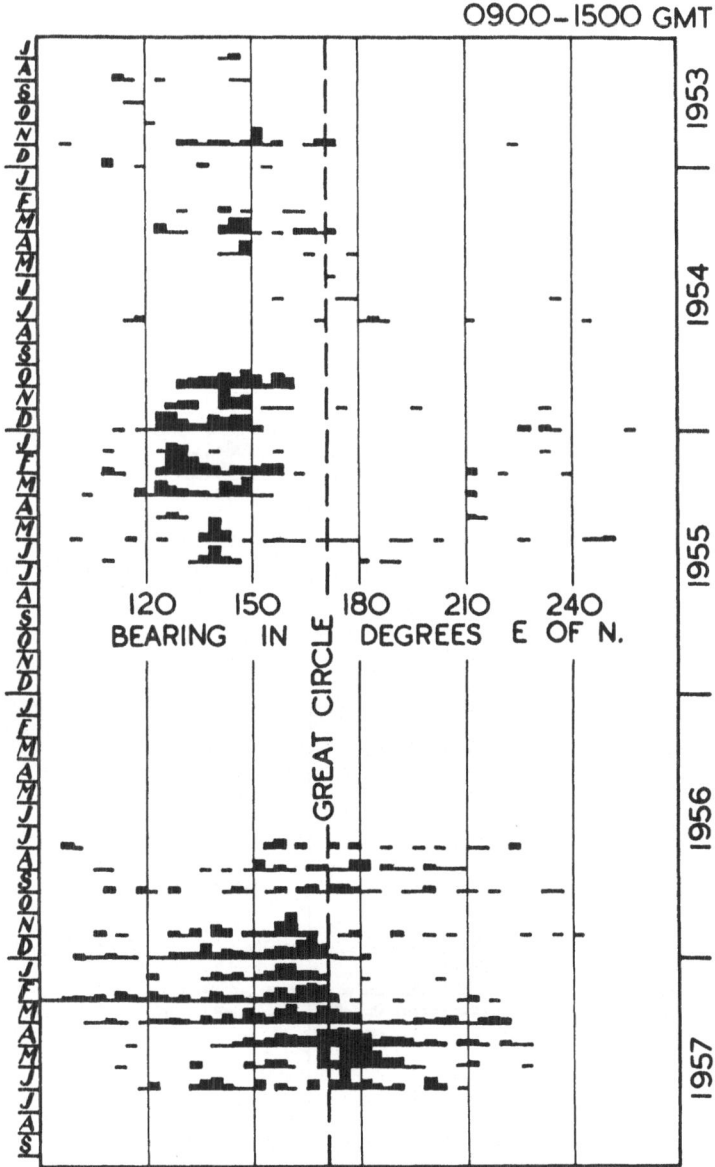

Fig. 10.2   Observed bearing angles over the years 1953-1957 during the 6-hour period 0900-1500 GMT.

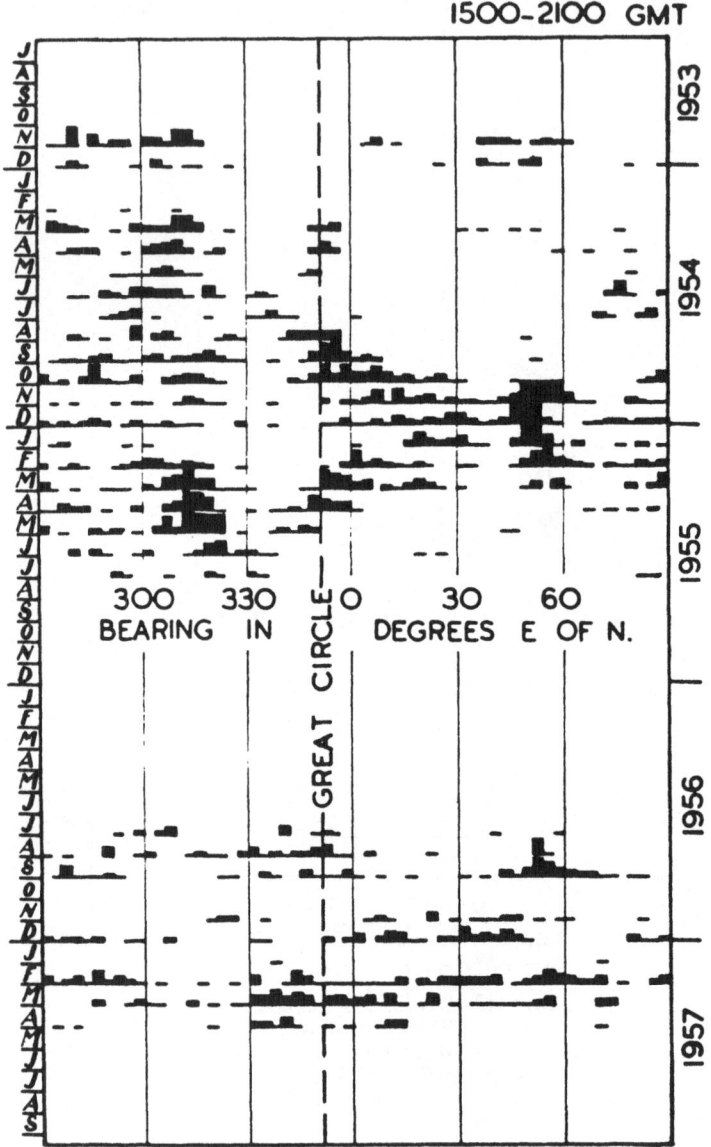

Fig. 10.3 Observed bearing angles over the years 1953-1957 during the 6-hour period 1500-2100 GMT.

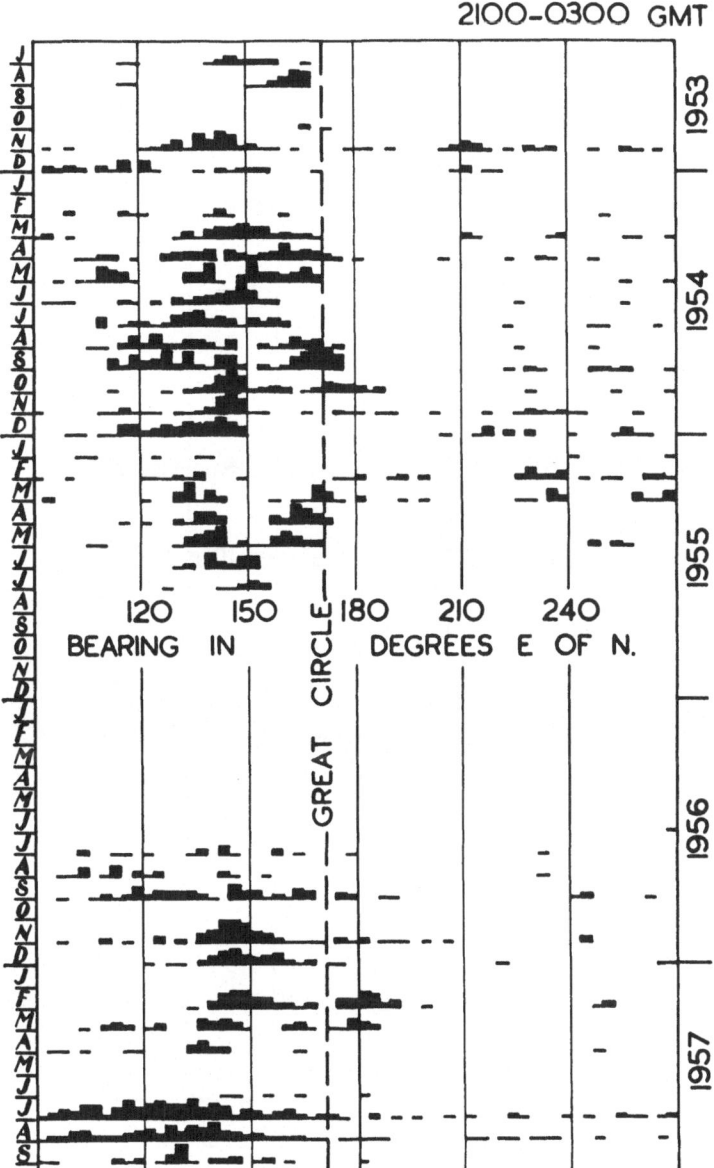

Fig. 10.4 Observed bearing angles over the years
1953-1957 during the 6-hour period
2100-0300 GMT.

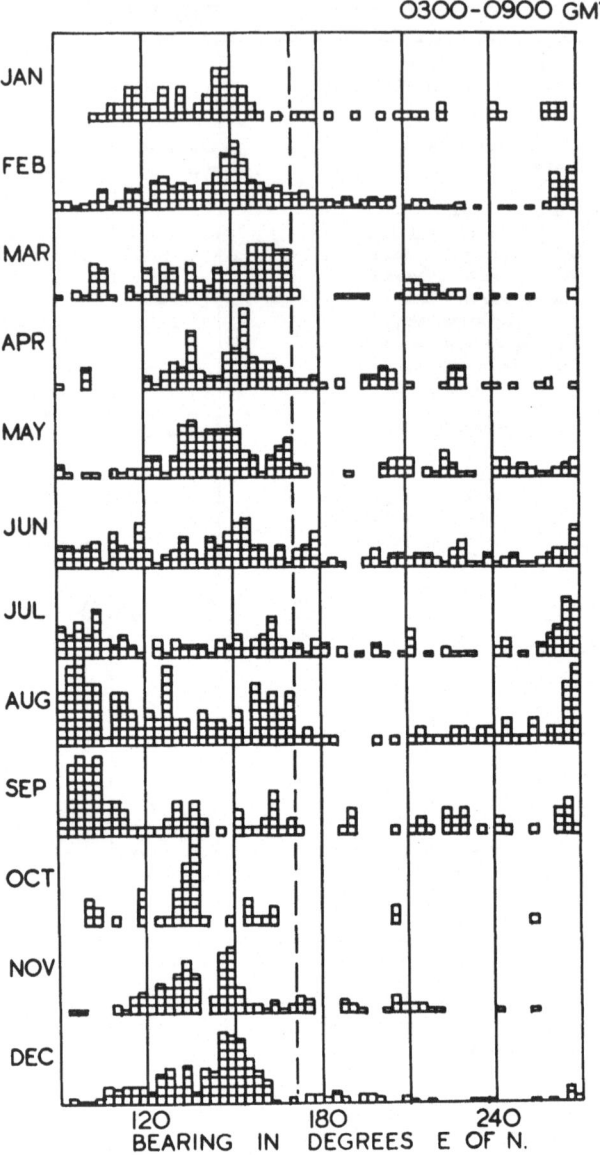

Fig. 10.5   Combined plot of observed bearing angles for
each month during the 6-hour period 0300-0900
GMT.

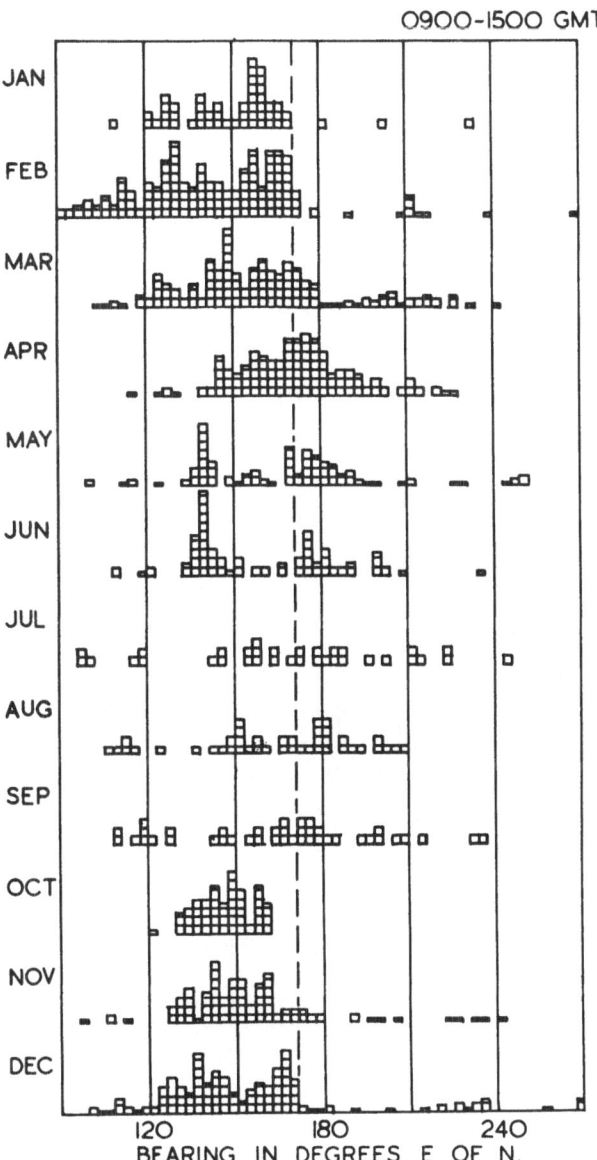

Fig. 10.6   Combined plot of observed bearing angles for each
month during the 6-hour period 0900-1500 GMT.

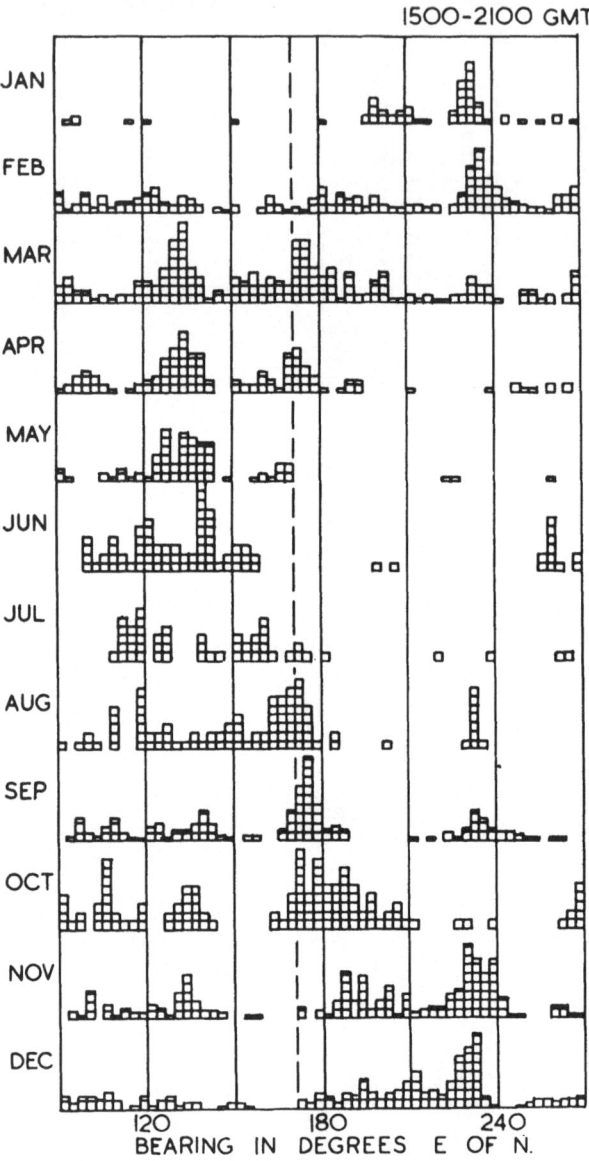

Fig. 10.7   Combined plot of observed bearing angles for
            each month during the 6-hour period 1500-2100
            GMT.

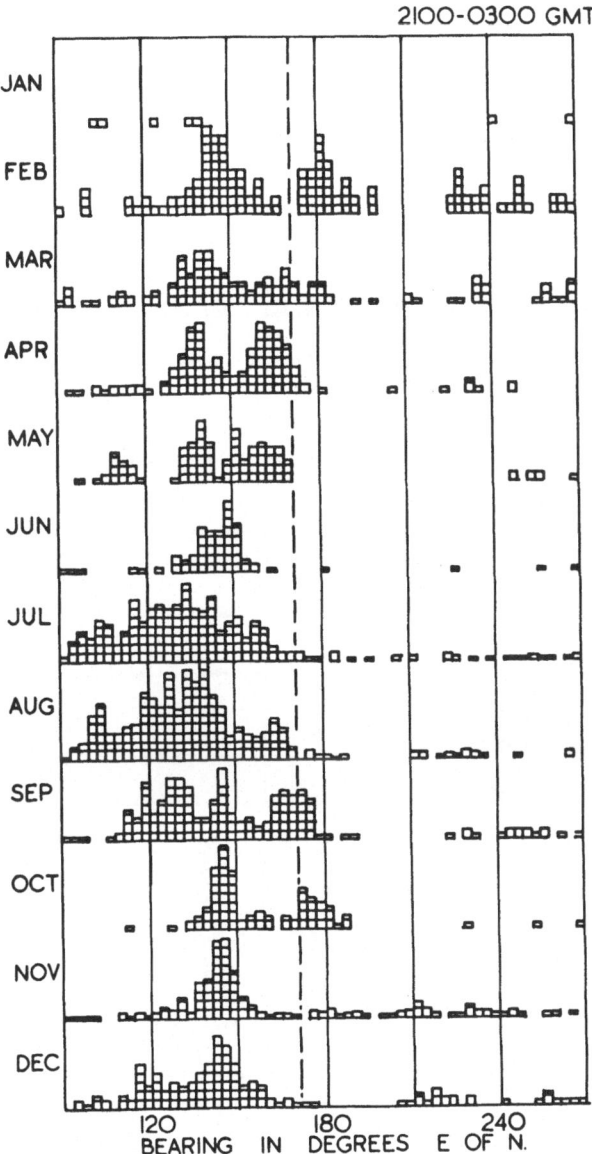

Fig. 10.8 Combined plot of observed bearing angles for each month during the 6-hour period 2100-0300 GMT.

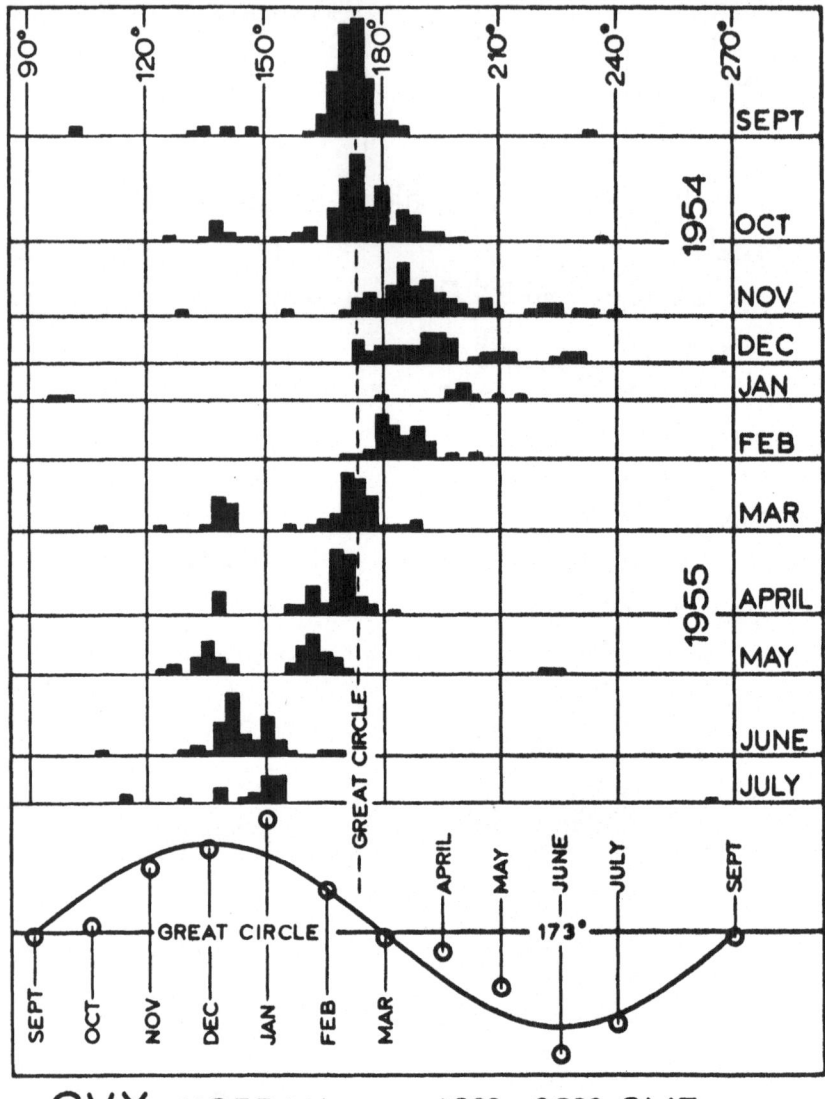

GVY 11955 kHz   18⁰⁰–23⁰⁰ GMT.

Fig. 10.9   Bearing angles for a particular station observed
             over a limited time interval for the months
             shown.   The lower diagram shows the variation of
             the average bearing with the season.

CHAPTER 11

SUDDEN BEARING CHANGES

Before leaving the subject of mean bearings (i.e. those from which the scintillation effects have been removed by an averaging process) we will mention briefly a phenomenon which appears to be quite common in those cases of near-antipodal propagation with which we are concerned here. In very many instances, it has been noticed that the received bearing changes by a large angle in a very short time. Changes of 30° in 2-3 minutes are not uncommon. Some examples of such changes are shown in fig. 11.1. These were not isolated examples but were very common during the years 1956-1957 when the bulk of these bearing measurements were obtained. The outstanding feature is that the jumps were always in the direction of decreasing bearing (notice that the time scale increases vertically in fig. 11.1). It is probable that these changes are associated with transmission through the auroral region in that they seem to indicate propagation "around" either one edge or another of an absorbing stop. It is probable that, under these circumstances, we can regard the overall propagation conditions over the bulk of the path as determining what portion of the auroral region is illuminated by transmitter, the final incoming direction being determined by the position of that edge which is most brightly illuminated, to use the optical terms.

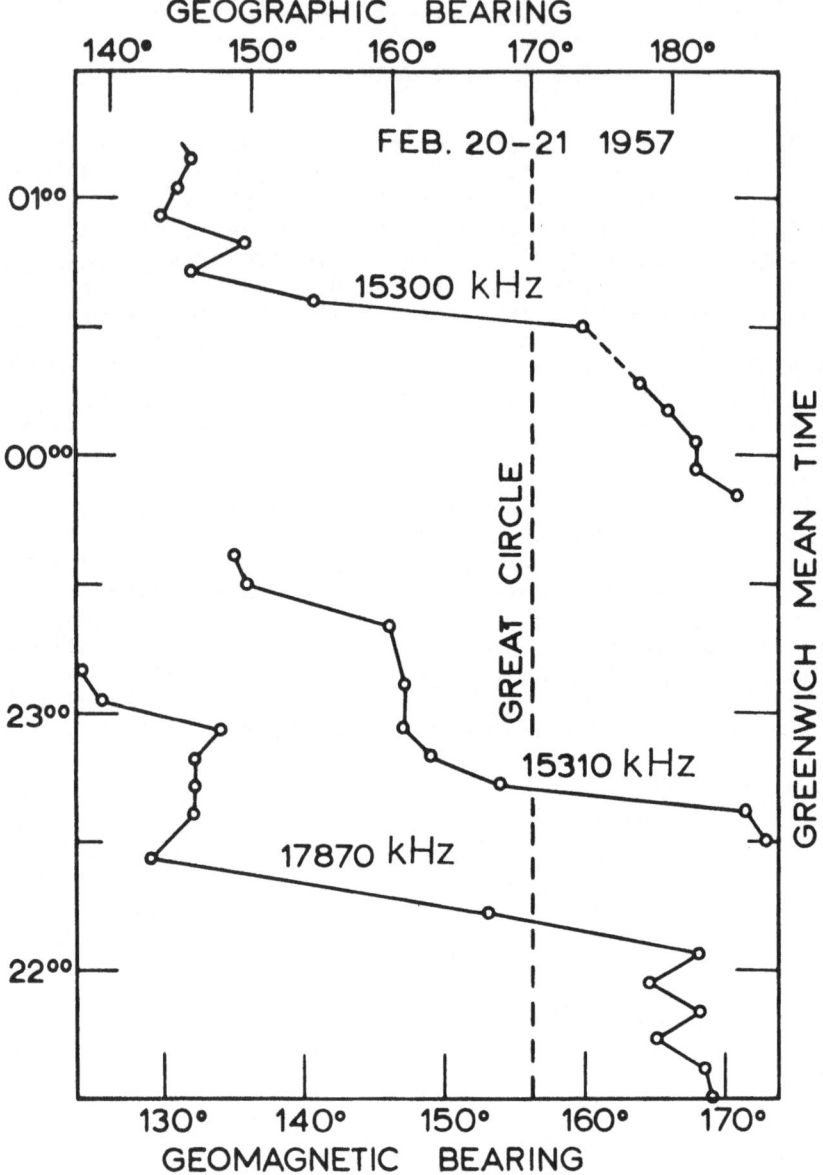

Fig. 11.1   *Typical examples of large sudden shifts in bearing. These are always from right to left in this type of representation. Note that time increases upwards.*

CHAPTER 12

VARIATIONS OF ELEVATION ANGLE

In spite of the obviously great difficulties in iso-
lating specific effects in the case of near-antipodal
propagation, we have seen that, in some instances (for
example, the results in fig. 10.9) it is possible to obtain
examples which indicate some of the mechanisms operating.
Another result of this nature was obtained during the
measurements of bearing angles in connection with the
effects of transmitter beaming directions which have
already been discussed (fig. 7.9). On this particular
occasion, the received bearings were clustered about a
mean bearing which was approximately the great circle bearing. This fact would, of itself, indicate that conditions
were quiet and that there was no very intense absorption
in the auroral region. In these experiments, the elevation
angles of the incoming waves were also measured; these
elevation angle measurements were much more significant
than in most experiments on very long distance paths since,
in this case, the transmissions were pulsed and it was
possible to obtain bearings and elevation angles of various
components in the received complex pulse. When the average
received elevation angle for each 3° range of bearing angle
is plotted as a function of this bearing angle, the results
shown in fig. 12.1 are obtained. It is noted that the elevation angle remains relatively constant at about 17° until
the bearing departs by about 15° from the great circle
direction. At this point the elevation angle increases
rapidly with increasing deviation of the bearing angle from
the great circle direction. Similar results have been
obtained on shorter distances but this is the only occasion in which the spread of the bearing angles was small
enough for this effect to be observed on near-antipodal
transmission.

The result can be explained quite simply of course since
the larger the deviation of the incoming direction from the
great circle direction, the greater the change in direction
required at each reflection point along the path or, alternatively, the greater the number of reflection points
required. The higher elevation angles immediately indicate
a larger-than-normal number of reflection points and hence
the possibility of greater deviation of the incoming
direction.

81

A simple treatment of this case in terms of small circle propagation paths leads to an estimate of the size of the antipodal area which agrees with what we deduced from fig. 7.3. As in that case we are considering here that particular antipodal area which is defined in terms of the probability distribution of the spread of directions of the time-averaged incoming bearings.

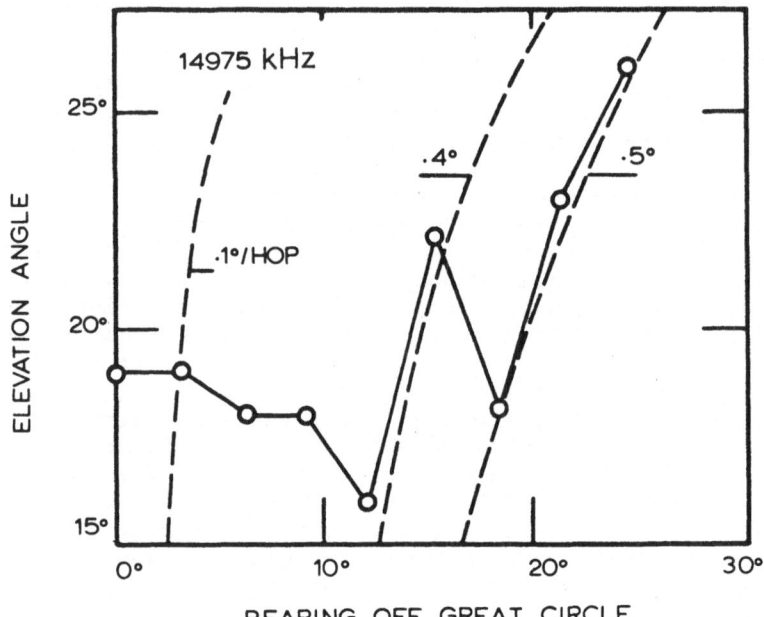

*Fig. 12.1 The increase in elevation angle of a received sig-
nal as the deviation from the great circle direction
increases. The dashed lines are identified by the
possible change in azimuthal direction of a ray at
each ionospheric reflection required to give the
plotted relation between elevation angle and deviation on a simple small-circle theory of propagation.*

The general appearance of a short transmitted pulse
(commonly 100μsecs long) when it is received at a distant
station is that of a number of overlapping pulses, all of
which apparently fade independently. Some typical shapes
are shown in fig. 12.2.b. It is extremely difficult to
find what may be termed fixed pulses within the groups.
Part of the reason for this is that the individual pulses
overlap and therefore produce interference effects between
themselves but it also seems that the relative positions
of the constituent pulses are changing continually. The
grouping of the pulses has been found to be usually in one
of three forms.   Either there is a single main group as
at A (fig. 12.2b.) or at B. In the example, A is the group
arriving approximately via the short great circle route and
B is that group which has travelled approximately along
the long great circle route. As a second form, both these
groups may be present simultaneously. As a third form,
there may be a single main group of pulses with a second
small pulse or group of pulses a few milliseconds later
(type C in fig. 12.2b). The origin or explanation of this
trailing pulse seems to vary from time to time as is
discussed below.

Measurements of directions of arrival of the pulse
group as a whole made, for example, using a rotating interferometer will give similar results to those made on a
continuous signal from the same source. However, if the
pulse repetition frequency is sufficiently stable, it is
possible to select parts of the pulse group and thus to
study these parts separately. Such a system is shown
as a block diagram in fig. 12.2a where the antennas are
those of the interferometer. With this system and with
the gates adjusted so as to position the first part of the
pulse in gate I, the main part (including the position of
the average maximum pulse) in gate II, the trailing edge
of the main pulse group in gate III, and any extra trailing
pulses in gate IV, results have been obtained as in fig.
12.3. Various groups of results have been classed together
to yield overall typical pictures of the characteristics
of the incoming pulse group. The two blocks labelled
rhombic and dipole refer to the transmitting antennas used
and, since these were switched at 10 minute intervals,
refer to slightly different time periods.

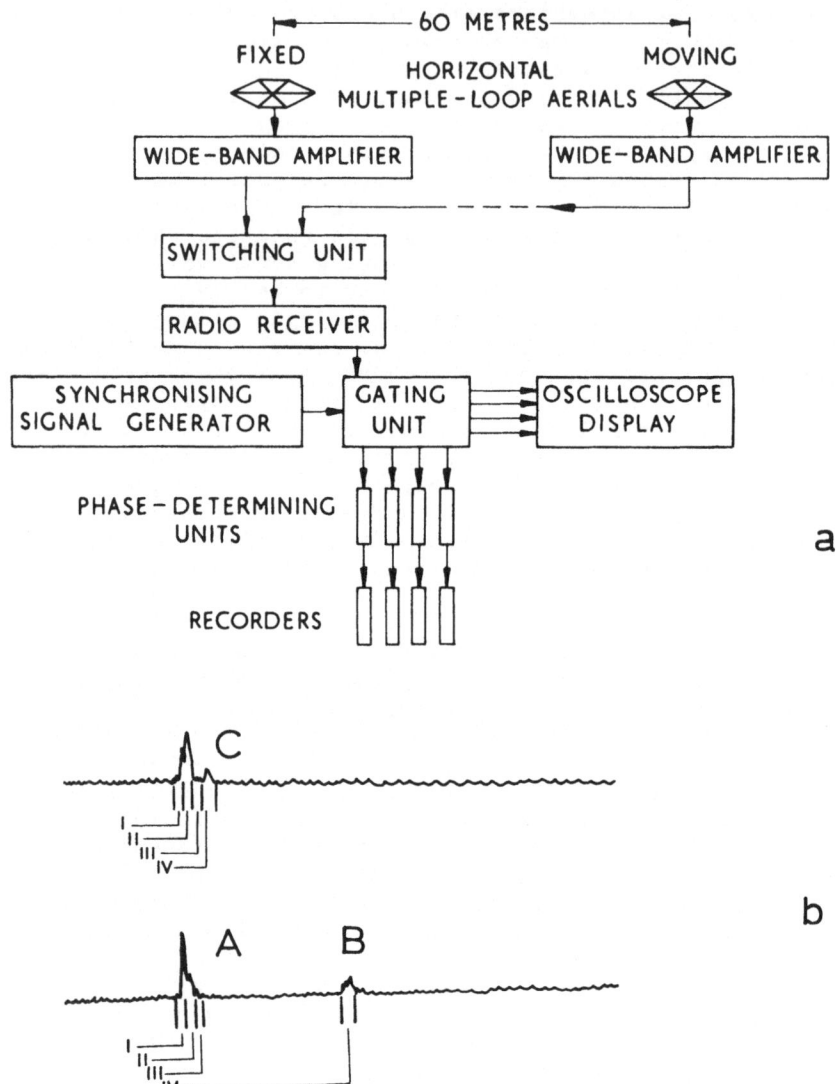

Fig. 12.2 Equipment for determining the angles of arrival of pulse groups such as those shown in the two lower diagrams.

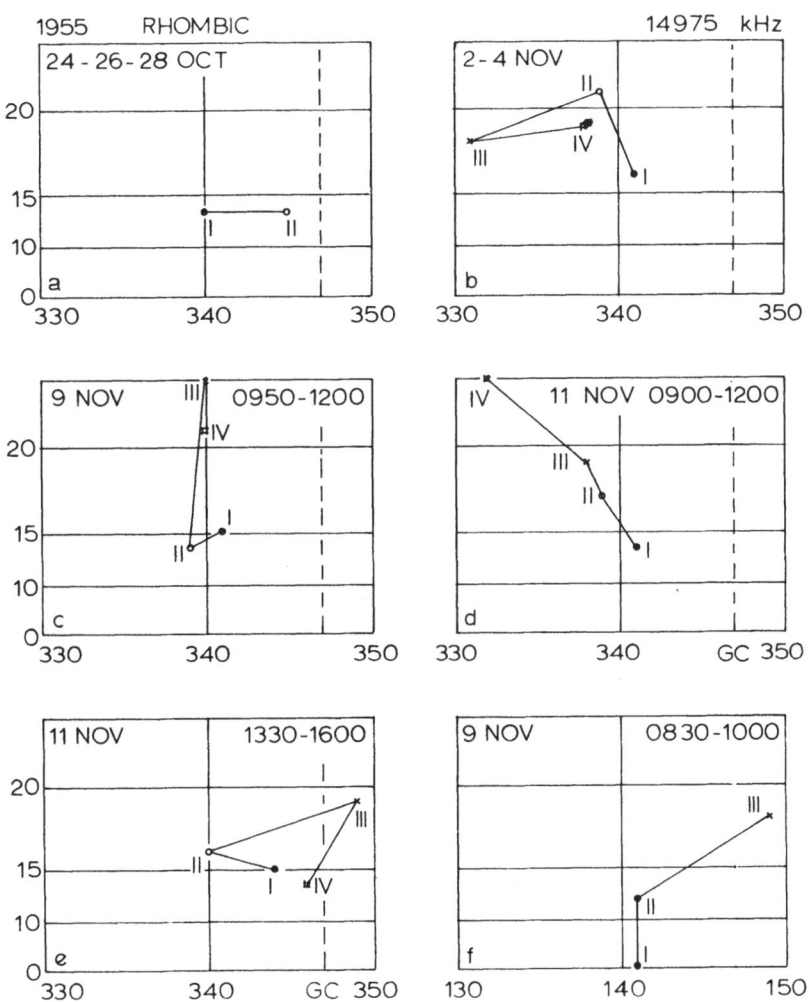

Fig. 12.3 (a to f) Angles of arrival of the pulse groups
specified in figure 12.2 when a rhombic
transmitting antenna was used.

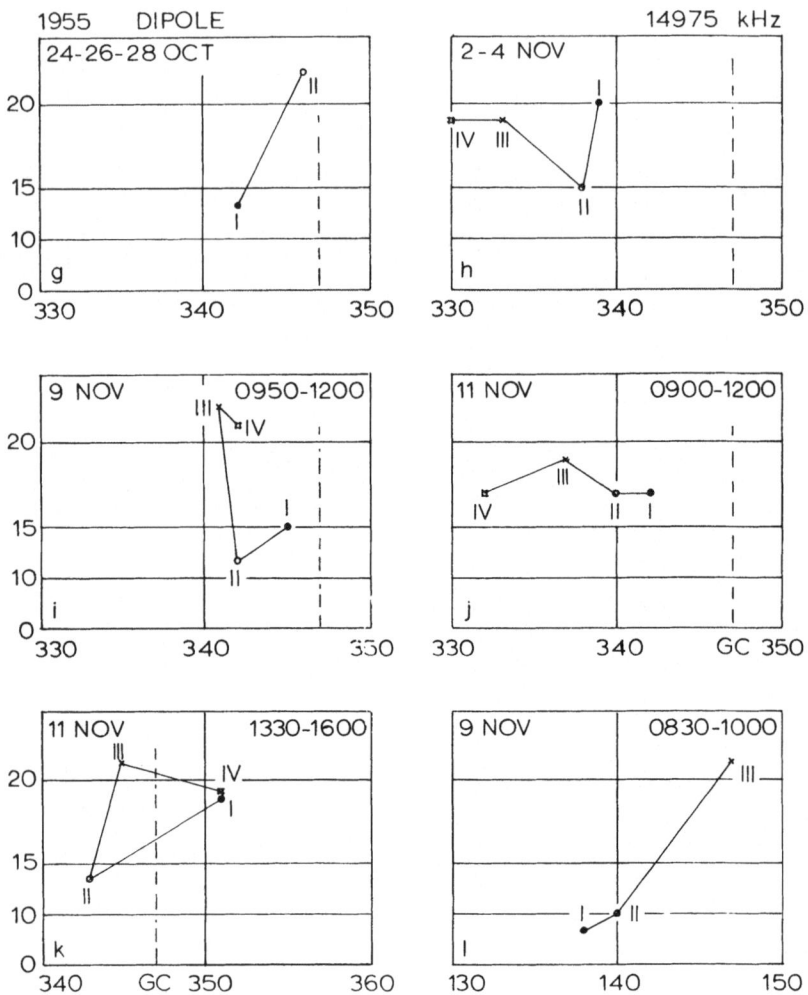

*Fig. 12.3 (g to l)   Angles of arrival of the pulse groups
specified in figure 12.2 when a dipole
transmitting antenna was used.*

In each diagram the bearing and elevation angle (averaged over about 4 or 5 results) for different parts of the
pulse group are plotted. Figs 12.3 f and l are for the
long path group (B in fig. 12.2). All the remainder are
for the short path group with the trailing pulse (c in fig.
12.2)   appearing as point IV (i.e. from gate number IV)
in b,c,d,e and h,i,j,k. The direction of arrival of the
trailing pulse (IV) is not significantly different from
those of the main group of pulses except in d and j.
This indicates that the significant delay in arrival of
the trailing pulse does not arise because it has travelled
in a different propagation mode from the rest of the pulse
group. It is probable that it represents energy which
has been reflected (possibly back-scattered) from some
particularly rough patch of the earth.

Most of the diagrams in fig. 12.3 show a general
tendency for the first part of the arriving pulse group
to be at a somewhat lower elevation angle than the remainder.
Further work has shown that this tendency persists but
certainly does not always occur; even from the relatively
few results shown in fig. 12.3 it is obvious that it is
misleading to regard the various pulses in a group to
have travelled by the same route but to have elevation
angles which increase with their delay. The mechanisms
by which the various delays occur are obviously very
complicated.

CHAPTER 13

THE SHORT-TERM CHARACTERISTICS

OF THE RECEIVED WAVE (SCINTILLATIONS)

In all of the above discussion we have been concerned with the averaged characteristics of the waves; in particular, most of the results have referred to the bearing angles averaged over periods of about half an hour or more. These considerations are important in deciding, for example, the direction in which a beam receiving antenna should be oriented so that it is capable of accepting the signal over the expected range of bearing deviation. Once this has been accomplished there is still the problem of the rapid variations of the incoming signal which arise largely from interference between various component waves. All such effects, which include the rapid fading of the amplitude of the signal and the rapid fluctuations of the wave-normal direction in both the horizontal (bearing) and vertical planes are classed together as scintillation effects.

We have noted previously that the classification which has been adopted in this work *viz.* the regular diurnal variation, the day-to-day deviations from this regular behaviour, and the rapid random interference effects which have been called scintillations, refers to different manifestations of the same basic physical process i.e. the refraction of waves in the ionosphere, the differences being mainly in their time scales. Since it is the *changes* in the various quantities that are of interest and since these changes are associated with movements of the ionosphere, this time scale may also be regarded as a size scale in the ionosphere. Thus the scintillations are associated with the smallest irregularities in the ionosphere. We note that, in the case of the variations arising from the larger scale irregularities, it is the directions of waves which are affected i.e. we are dealing with overall refraction of the wave, but, in the case of the scintillations, the irregularities are so small that different parts of the wave are refracted by different amounts so that a whole series of wavelets travelling in different directions arrive at the receiver. Under these circumstances, the changing wavelets can and do interfere so that at some times they will add constructively to give high signal strengths at the receiver and at other times they will add destructively (because of phase changes) to produce low signal strengths

at the receiver. We thus have amplitude variations
associated with changes in instantaneous bearing directions in the case of the scintillations. A very large
amount of work has been carried out on the general
problem of deducing the characteristics of the incoming
cone of rays from the statistics obtained by comparing
the signals received on various pairs of spaced aerials.
This work has been carried out largely in the fields
of radio astronomy and in the investigation of ionospheric winds which may be detected from the way in
which long-lived irregularities drift horizontally in
the ionosphere.

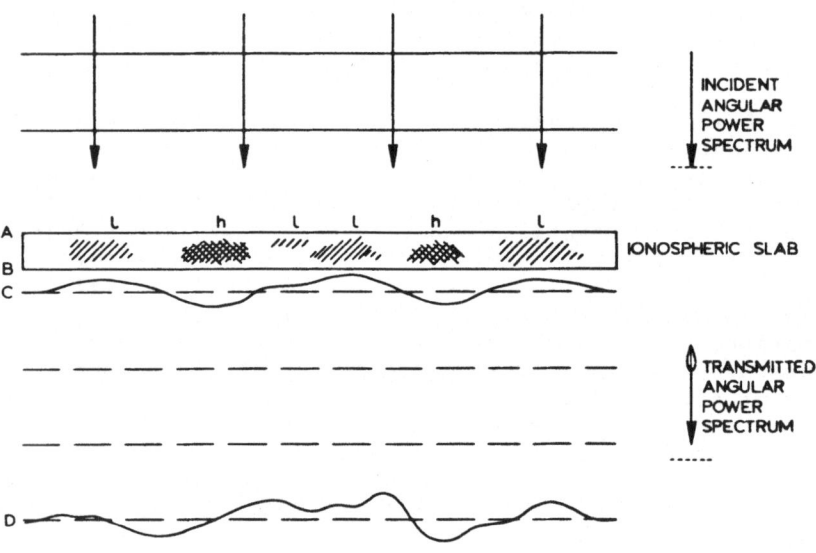

Fig. 13.1   The distortion of the wavefront as a plane wave
travels through an inhomogeneous medium AB.   The
shape of the resultant wavefront continues to
alter as it travels from C to D.

The foundation of the theory applicable to the scattering of waves by irregularities within the ionosphere was laid
down by RICE in his classical work on the theory of random
noise (RICE, 1944, 1945). The extension of this theory to
the ionospheric case is a process which has occurred over a
period of many years. The theory is particularly straightforward if the irregularities in the ionosphere are regarded
as *opaque* "blobs" and a great deal of work in this field has
been carried out on this assumption. In actual fact, of
course, the irregularities consist of rather small departures
of the electron density from the average or background value,
so that the blobs in the ionosphere are not opaque but are
regions in which the refractive index is only slightly different from the average value. Such a medium is sometimes referred to as being *diaphanous*. This means that various portions
of a wave traversing the medium undergo different phase delays i.e. appear to have travelled different distances in the
ionosphere. This may be illustrated as in fig. 13.1 where we
have shown a plane wave incident on a slab of ionosphere AB
from above. In this slab there are regions of low electron
density (l) and regions of high electron density (h). The
plane wave-fronts which entered the region from above can be
pictured like the wave-front C immediately after leaving the
lower boundary of the ionospheric slab. This corrugated wave-
front represents a rather complicated wave; it is conveniently described in terms of its *angular power spectrum* which is
actually a type of polar diagram of the directions in which
the power in the wave is travelling. For the incident wave,
since all the power is travelling vertically downwards, the
angular power spectrum is simply a downward pointing narrow
spike. The emergent wave contains two main components.
Firstly, there is the component which may be regarded as the
undeviated part of the incident wave which must be present
since, although the wave-fronts are corrugated, they are on
the whole travelling in the same direction as the incident
wave. Secondly, there is the part which accounts for the
corrugations themselves since these occur as interference effects between the various waves travelling in a range of directions which are clustered about the undeviated or *specular*
direction. The two different angular power spectra are shown
in the right-hand part of fig. 13.1. Since there is no absorption in the ionospheric slab, the total power in the
transmitted wave will be the same as that in the wave incident on the screen. In the incident angular power spectrum,
this total power is represented by the length of the line

which shows the direction of power flow.   In the transmitted angular power spectrum, the power flow still
remaining in this direction is again represented by the
length of the arrowed line but, since some power is
scattered in other directions this length is now smaller.
The total scattered power is represented by the area under
the curve which depicts the way in which the scattered
power is spread out in various direction.

       The plane wave incident on the ionospheric slab is
assumed to have the same amplitude at all positions of the
wave-front.   The irregularities in the refractive index
of the ionosphere will, in the process of introducing phase
changes in this wave act to some extent as weak lenses
(converging when the irregularity consists of a deficit of
electrons and diverging when the irregularity consists of an
an excess of electrons).   It is a fairly safe assumption
that, in most cases, these lenses are so weak that there is
no appreciable change in the amplitude of the wave by the
time it has emerged from the bottom of the ionosphere i.e.
the wave at C (in fig. 13.1.) has variations in the position
of its wavefront but negligible variation in its amplitude.
As the wave progresses downwards, there will arise, however,
variations in the amplitude of the wave as the various
component wavelets which form this irregular wave and which
are travelling in slightly different directions get in and
out of phase with each other.   Finally, at a sufficiently
great distance below the ionosphere, the amplitude variations
will have built up to a stable level (statistically speaking).
This way in which this buildup of the amplitude variations
occurs depends on the range of sizes of irregularities in
the ionosphere and on the relative number of irregularities
in each small size range.   We will not consider this
problem here, mainly because there does not appear to exist
a completely satisfactory solution.   We will, however, be
concerned with the variations in amplitude and with the
size of the corrugations in the wavefront when these have
reached a state which can be regarded as approaching the
fully developed condition.   This state of affairs is
referred to by saying that we are considering the *far-
field region* of the screen.

As the corrugated wavefronts progress downwards they will not retain their original shape since their component waves, travelling in slightly different directions, will have different phase relationships at different distances below the ionosphere. There will, however, be some statistical property of these waves which is related to the shape of the corrugations in the wavefront just after it left the ionosphere and hence to the irregularities in the ionosphere which gave rise to these corrugations.

In this section an attempt will be made to present the main characteristics of such scintillating signals. This could be carried out merely as a classification of the results which have been observed but, since this does not make it possible to predict what will happen under a new set of circumstances, the results will generally be discussed as manifestations of events occurring in the ionosphere. The investigation of the characteristics of irregularities in the ionosphere is a subject which is still very active and the general picture of their dis-tribution in space and their range of sizes is far from complete. A technique which is now widespread is to use the signals being transmitted from orbiting satellites. The method is very satisfactory for the relatively large irregularities (which give rise to effects which are mainly refraction effects) but is not quite as satisfactory for investigating the small irregularities which give rise to scintillations. The reason for this is that the satellite is moving so rapidly that the pattern on the ground changes too quickly for the rapid variations to be observed and recorded. However, techniques are improving quite rapidly and this problem does not seem to be insuperable. In the meantime, we can make reasonable estimates of the irregular structure of the ionosphere from the data which is available. We have available, and will discuss, the results of many investigations made on actual propagation paths so that, even though these results can not be related accurately to the structure of the ionosphere at the present time, they do provide a guide to the effects which are occurring. Even when considerably more information on small ionospheric irregularities is available, it may well turn out, as in the case of the day-to-day deviations of the bearing direc-tions, that their occurrence is essentially random and hence not predictable.

Some results which indicate the sizes of irregularities to be expected are shown in fig. 13.2* in which the number of observed irregularities is plotted as a function of the size. In general, it is found that the number density is inversely proportional to the spatial extent (size) of the irregularity. In fig. 13.3* the distribution in height of rather large irregularities is plotted.

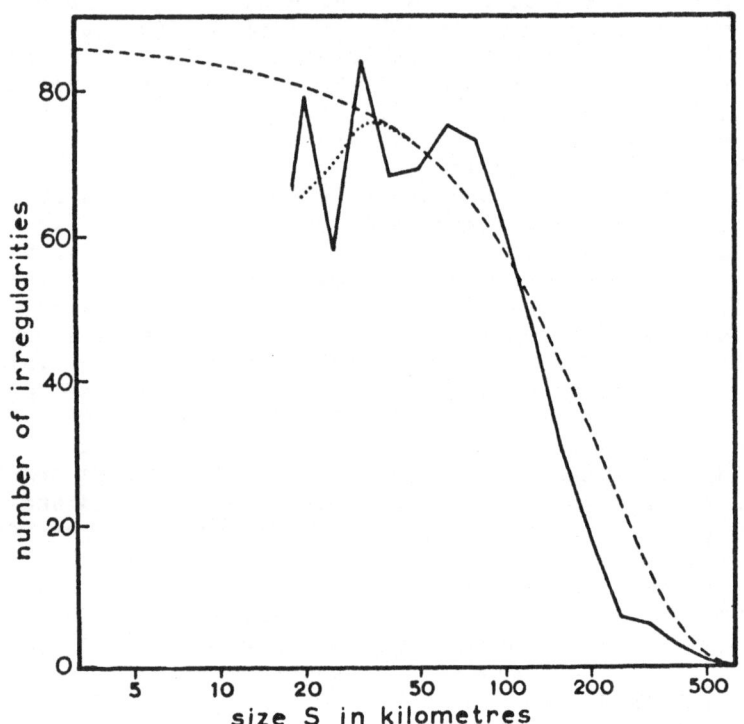

Fig. 13.2   Experimental curve showing the way in which the
            number of irregularities of a given size changes
            with the size.

*I am indebted to Dr J.E. Titheridge and Dr G.F. Stuart for these results.

These heights were obtained by observing the effects
of the same irregularity on radio signals from a satellite
at two different observing sites on the ground that were
separated by about 100 Km.   The ratio of the total electron
content (C) to the size (S) of the irregularities is
plotted in fig. 13.3 as a function of the height. This
ratio is roughly proportional to the background electron
density (from the experimental results) so that it is
concluded that irregularities of intensity proportional
to the background electron density occur fairly evenly
distributed throughout the ionosphere with their relative
number inversely proportional to their size i.e. there are
many more small irregularities than there are large ones.

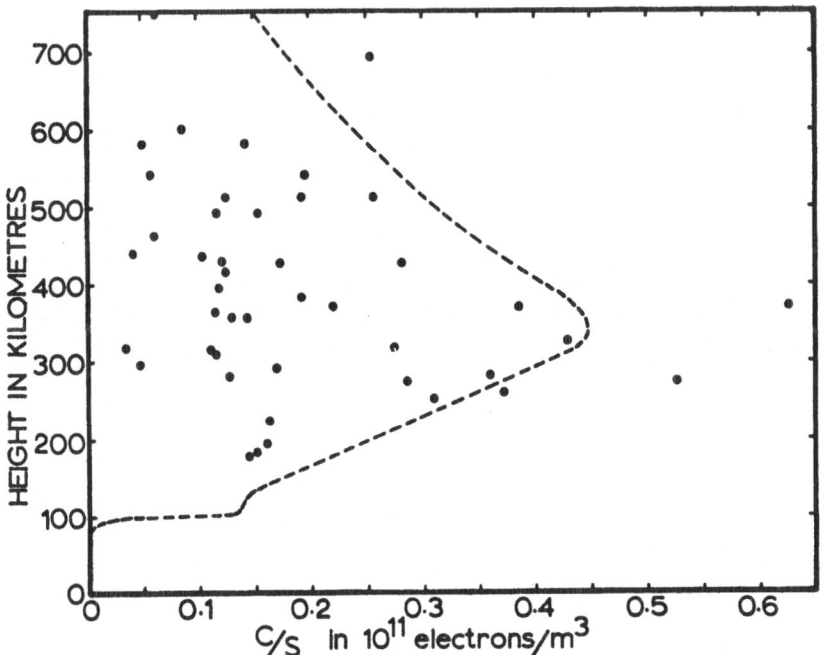

Fig. 13.3   The way in which the density of an irregularity
            depends on the density of the part of the iono-
            sphere in which it occurs. The dashed line is 10%
            of the average ionosphere at the time of the
            observations.

Since these results refer to a particular location, we can not deduce that the same behaviour occurs everywhere, but we can make the important deduction that the irregularities do occur throughout the ionosphere. This means that we do not have to consider the propagation mode in detail in long-distance transmission paths but can assume that irregularities are encountered at each reflection from the ionosphere.

In order to describe these electron density irregularities in terms of their effect on a plane wave passing through the ionosphere, it is necessary, firstly, to derive an expression for the angular power spectrum of the wave leaving the ionosphere. The relevant characteristics of the ionosphere are

(i)   the amount of phase change introduced by the irregularities, conveniently specified in terms of the RMS phase deviation $\overline{\phi^2}$ averaged over the whole wave front,

(ii)   the size of the structure of the irregularities, which depends on both the size of the individual irregularities and the spacing between them and may be specified in terms of the average correlation between the phase change at any point on the emergent wave-front and that at some other point distant x from it. This correlation function is called the *autocorrelation function of the phase deviations* and may be written, in a general form as r(x). This is a curve which essentially specifies how rapidly the phase difference between two points on the emergent wave changes as we take these two points and separate them by the distance x. If the two points are coincident (x=0) then r(0) = 1 since there is obviously no difference in phase between one point and itself i.e. there is complete correlation. If the two points are very far apart so that there is no dependence of the phase at one point on that at the other, r(x) → 0 i.e. the two points are completely uncorrelated. If r(x) drops to zero only slowly as x is increased, the irregularities are mainly of large size but if r(x) drops to zero very rapidly as x is increased, the irregularities will include a large number that are of very small dimensions.

We note that the two parameters given above *viz* $\overline{\phi^2}$ and the function r(x) are essentially properties of the irregularities in the ionosphere. They do depend, however on the frequency of the wave which is traversing the region concerned since the phase change introduced into a wave by

a given amount of electron density deficit or excess
decreases as the frequency increases.

These two parameters may be conveniently combined to
give another function of basic importance in this theory
which is called the *generalised autocorrelation function*
of the screen. This function can be computed readily from
the wave-field just after it has left the screen but it is
not restricted to the field just as it leaves the screen,
where there is as yet no amplitude variation arising from
interference effects but only the phase variations introduced by the screen. It was shown by BRAMLEY (1955) that
the generalized autocorrelation function $\rho(x)$ is given by

$$\rho(x) = \exp - \left\{ \overline{\phi^2} \, (1 - r(x)) \right\}.$$

In this expression, the correlation function $r(x)$ is a
property of the phase variations in the screen and $\overline{\phi^2}$ is
the variance of these same phase variations i.e. the function
is determined completely by the phase characteristics of
the screen. At a distance from the screen, the amplitude
of the wavefield will not be always unity as it is at the
screen so that some different expression will be expected.
In fact if the amplitudes and phases of the wavefield at
two points a distance x apart are given by a_1, a_2 and ψ_1, ψ_2
then the generalized autocorrelation function is given by

$$\rho(x) = \left\langle \, a_1 a_2 \, \cos \, (\psi_1 - \psi_2) \, \right\rangle$$

where the diagonal brackets indicate that the quantity
within the brackets has to be averaged over all positions
within the wavefield at the distance from the screen which
has been chosen.

Since this function is the same for any section through
the wavefield, it may be determined at the ground, thus
yielding direct statistical information on the extent and
structure of the phase changes introduced by the ionosphere.
A representative generalized autocorrelation function is
shown in fig. 16.2.

There is a well-known mathematical relationship between
this generalized autocorrelation function and the angular
power spectrum which is of great value in interpreting
measurements on the scintillation characteristics of a radio
wave. If either of these functions is found, the other can
be obtained from it.

If we write $P(\theta)$ for the angular power spectrum, we have (after BOOKER, RATCLIFFE and SHINN, 1950) that

$\rho(x)$ is the Fourier transform of $P(\theta)$ and hence,

$P(\theta)$ is the Fourier transform of $\rho(x)$, also.

It is convenient, when considering effects occurring at a single antenna (which may, however, consist of a number of elements) to think in terms of the angular power spectrum. When one is concerned with systems made up of a collection of individual antennas spaced along the ground, as in diversity systems, it is then convenient to think in terms of the generalized autocorrelation function.

In the cases which are considered in succeeding sections, we will be concerned with long-distance propagation paths so that about half the reflection points will be at the surface of the earth. In the frequency range with which we are concerned the earth may be regarded as being a good reflector but the effects of its roughness must be taken into account. If we consider the simple case of a plane wave which is vertically incident on the earth's surface as in fig. 13.4. we see that the irregularities in the surface produce relative phase changes between the various parts of the reflected wave. The reflected wavefront C will have the shape shown.

If, again, we assume that the amplitude changes arising from the focussing and defocussing effects of the curved

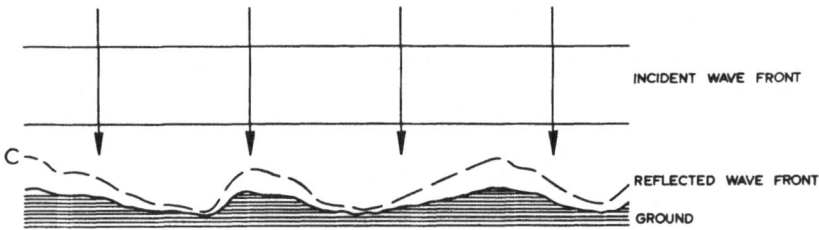

Fig. 13.4   The way in which a ground reflection changes the shape of the wavefront showing that a rough ground can be regarded as a phase-changing screen.

mirrors which are formed by the irregularities of the
ground are negligible close to the ground, the wavefront of
the reflected wave just as it leaves the ground may be
described by a generalized autocorrelation function of
exactly the same form as that which applies to the ionosphere.

In comparing ground and ionospheric reflections, we
may note that the ionosphere may introduce perhaps one or
two complete 360° rotations of phase angle difference
between various points of the wave which is reflected from
it when we consider a practical propagation case, but the
ground reflections will cover a very much wider range of
conditions. For example, let us take 20 MHz (15m) as a
typical operating frequency. The ground will appear smooth
if the height fluctuations from the mean level are less than
± 4m (the Rayleigh λ/4 criterion). There are many places
on the earth's surface, including relatively calm seas
where this condition will hold. At the other extreme, we
have places where the height variations are many hundreds
of metres in short distances. Such places will be extremely rough compared with the ionosphere since a variation
of ± 100m will give ± 7 complete 360° rotations of the phase
angle. We may also note that, whereas the ionosphere and
the sea surface are reflecting regions which are changing
with time, the solid earth surfaces are generally not
changing, the effects of wind on vegetation being generally
negligible at these frequencies. Since we are usually
concerned with the case where the first reflection point
is at the ionosphere we need not consider the fact that
some of the later reflection points are stationary since
they will be illuminated by a changing field and will thus
not usually give rise to steady scattered fields in any
particular direction. We may note, however, in passing,
that waves on the sea very often have a periodic structure.
Such a reflecting surface will give rise to a reflected
radio wave which travels in preferred directions i.e. the
angular power spectrum has distinct lobes. It is possible
that this effect could be important over short paths where
only one sea reflection occurs but it does not appear
likely that the effect would be important when many such
reflections occur. In Chapter 17, when the effect of many
reflections over a long path is considered in relation to
the formation of an antipodal focussing area, the assumption
will be made that all the reflecting regions can be described in terms of some type of average roughness.

CHAPTER 14

RELATIONS BETWEEN THE BEARING AND THE AMPLITUDE

In the preceding section, the scattering of waves by irregularities in the ionosphere was discussed mainly in terms of one passage of an initially plane wave through the ionosphere. Under the conditions in which we are most interested here, i.e. long-distance radio propagation, many encounters with the ionosphere will have occurred. Although none of these will be a passage right through the ionosphere, most being a double passage at an oblique angle through that part of the ionosphere which lies below the reflection level of the wave, the picture of an incident plane wave becoming a plane wave together with a series of scattered components will still hold. The scattered components will interfere and produce changes in the amplitude of the wave received at the ground, so that the received wave exhibits two types of fluctuation. The first of these, which arises from the corrugation of the wave front (the phase fluctuations) can be investigated if a direction-finder which can observe rapid fluctuations in the wave-normal direction is employed. The second type of fluctuation is the amplitude fading of the wave. Since both effects arise from the interference of the various waves which make up the angular power spectrum, it would be expected that there would be some statistical relationship between the amplitude and the wave-normal direction at any particular time. That this is a reasonable expectation can be seen by considering the following very simple example.

It is commonly the case, especially in relatively short-distance propagation, that the incoming signal contains two main components. These could be, for example, the wave arriving by a two-hop mode and that arriving by a three-hop mode. Let us ignore the fact that each of these will be made up of scattered components as well as the specular component and consider only the interference between the two specular components. (We may do this since, for a short path, the power contained in the specular component may be many times that in the scattered components; some actual measurements of this ratio are described later in this section). In general the specular components will arrive from different directions. Certainly, they will

have different elevation angles and if there is any lateral
tilting of the ionosphere their bearing angles will also
be slightly different. The interference pattern on the
ground will then contain both corrugations of the resultant
wave front (in this case, since there are only two interfering components, the corrugations will be quite regular)
and variations in the resultant amplitude. If the two
waves are of nearly the same initial amplitude, the resultant amplitude at the places where they happen to be exactly
out of phase will be very small. At the same time, these
will be the places where the wave-normal direction shows
the greatest deviations from its average direction. This
large deviation will be to either higher or lower bearing
angles depending on which constituent wave happens to have
the greater amplitude at the moment when they have 180°
phase difference as was discussed previously in connection
with fig. 3.2

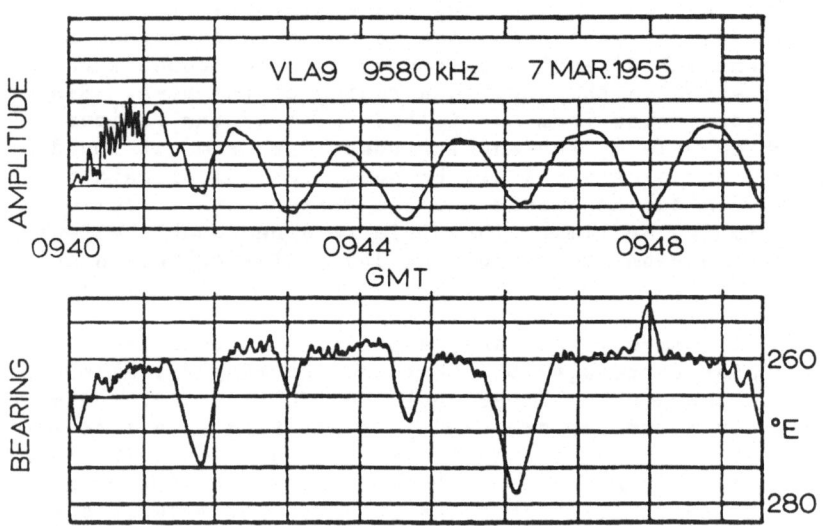

Fig. 14.1   Experimental record of amplitude and bearing
            fluctuations showing the field strength minima
            associated with large bearing excursions.

The direction of the apparent bearing deviation at times of low resultant amplitude depends on which of the two component waves happen to be the greater at these times. A small change in amplitude of one of the waves can thus have a very large effect on the apparent bearing (wave-normal direction). An experimental result showing this effect is reproduced in fig. 14.1.   It will be seen, from this record, that, at times of amplitude minimum, the bearing angle has shown large deviations to both sides of the mean direction on different occasions.

These curves, which are a *continuous* record of bearing and amplitude are obviously obtained with equipment which is different from the rotating interferometers which were used to obtain the results which have been employed for most of the work connected with the day-to-day variations of bearing angle. In fact, this instrument was developed so that this particular type of continuous recording could be made. The actual direction-finder is described in the literature (WHALE and ROSS, 1956).

In general, where a whole range of incoming angles is involved and not just two discrete angles as in the above example, the relationships between bearing and amplitude will not be as marked but there will usually still be some kind of correlation between the two quantities. The shape of the statistical distribution curves relating the bearing to the amplitude could be important in some types of receiving installation. For example, if the extreme bearing fluctuations are always associated with very low signals, it may be advantageous in some systems to design a receiving system which eliminates these extreme portions of the incoming range of signals.

Returning to the picture of the incoming waves as making up an angular power spectrum of constituent waves, it is seen that, if the scattered components are always distributed in a curve which has a particular shape, the whole distribution can be specified in terms of two parameters. These are

(i)   the ratio of power in the specular component to that in the scattered components which is equivalent to a signal-to-noise ratio in the better known context of circuit theory. In this case it is more descriptive to use the optical terminology and call this quantity the *coherence ratio* which is commonly denoted by the symbol B (although some authors have used b^2 for the same quantity).

(ii) the angular width of the distribution of the scatterd components which is conveniently specified in terms of the square root of the variance of the distribution, usually called the standard deviation of the distribution.

The total area N (for *noise*) under the curve representing the distribution of scattered components is the total power in these components and is known since, if we write

S = power in the specular component,

N = power in the scattered component,

then   S/N = B.

Normalize so that the total power is unity.

i.e.   S + N = 1

Then N = 1/(1 + B), and this, with a knowledge of the shape of the distribution, specifies it completely.

After only one passage through the ionosphere, a wave could still contain a very high proportion of specular component. The exact value will depend on the frequency, being higher for the higher frequencies, and on the particular circumstances but a relatively high value would be B = 10. At each successive passage through the ionosphere, more of the specular component is converted into scattered components. From the expression given in the previous section for the generalized autocorrelation function, it may be shown that

$$B = \frac{\exp(-\overline{\phi^2})}{1 - \exp(-\overline{\phi^2})} \qquad (14.1)$$

where $\overline{\phi^2}$ is the variance of the phase deviations introduced into the wavefront by the irregularities in the ionosphere.

If all passages through the ionosphere are identical then for n encounters with the ionosphere, the total variance of the phase deviations introduced into the wavefront will be about n times that for one encounter so that

$$B(n) = \frac{\exp(-\,n\overline{\phi^2})}{1 - \exp(-\,n\overline{\phi^2})} \qquad \ldots\ldots \ (14.2)$$

where the mean square phase deviation $\overline{\phi^2}$ refers to one passage only. As an example, fig. 14.2 shows the value of B which would be observed after various numbers of ionospheric reflections if $\overline{\phi^2} = .1$. This corresponds to B = 9.5 for one reflection only. It can be seen that B decreases very rapidly - after only 7 reflections the total power in the scattered components is equal to that in the specular component i.e. B = 1. The problem of determining B experimentally is left till a later section; it is noted here only that, for very long distances, B is likely to be quite small.

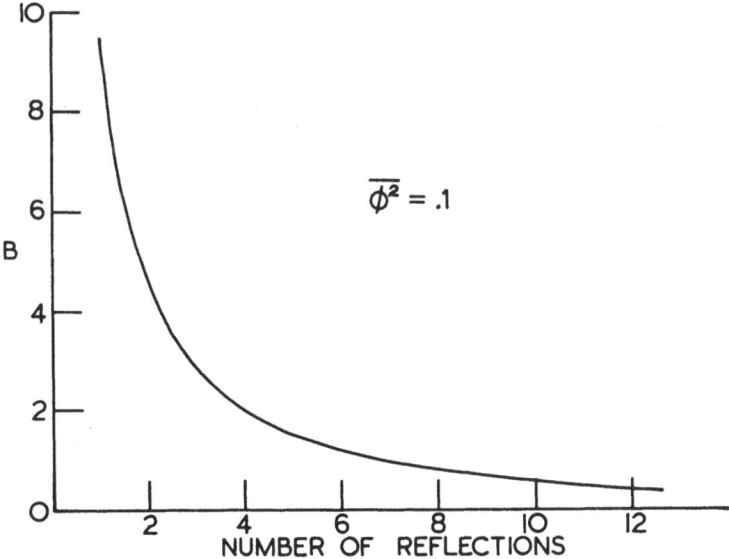

Fig. 14.2   The coherence ratio B as a function of the number of "steps" in the propagation path. The mean square phase deviation introduced by an average earth or ionospheric reflection is taken to be ·1 radian².

In fact,the situation has been oversimplified because, in propagation through the ionosphere, there will always be at least two apparently specular components present *viz.* the ordinary and the extraordinary components. Because of the presence of some large-scale irregularities i.e. irregulari- ties with large physical dimensions, some refraction of the wave will also occur. How, since a specular component is essentially a plane wave of absolutely constant direction and strength, the presence of a second component which interferes with the first and thus changes the resultant amplitude and even a small degree of refraction giving a change in direction means that a true specular component does not exist. The angular power spectrum drawn in fig. 13.1 (lower part of diagram) should thus consist of a very narrow sharply peaked group of *pseudo-specular* components together with the existing rather widely spread scattered components. Whether this group of pseudo-specular components will be resolved into its components in any actual measurement depends essentially on the aperture of the measuring system i.e. on the maximum separation between the various antennas which are used. A small-aperture system was employed to obtain the results presented in this section so that the pseudo-specular components will all appear to come from essentially the same direction. However, these components will still interfere, one with the other, so that the amplitude of the apparent specular component will change i.e. the value of the apparent B will not be constant. Such a variation will have the effect of reducing any measured ocrrelations between amplitude and instantaneous bearing fluctuations.

The statistical relationship between the amplitude and the wave-normal direction can best be presented in the form of two-dimensional or joint probability diagrams. These are essentially contour diagrams of a two-dimensional surface in which the probability contours are plotted on a co-ordin- ate system in which relative amplitude is plotted along one axis and wave-normal direction in the other. Each of these quantities is fading in its own characteristic way; for example, there are the records in fig. 14.1 which have already been discussed. Records which are more typical of the type of fading usually observed are reproduced in fig. 16.1 (lower two sections). Imagine that a pointer moves horizontally with the bearing fluctuations and vertically with the amplitude fluctuations. Then the relative time it spends within the boundaries of any specified small square on the board is a measure of the two-dimensional probability in that region.

The first two diagrams, figs. 14.3 and 14.4, are plotted for B = 0 i.e.there is no specular component. The horizontal scale is in units of the sine of the deviation of the wave-normal direction while the spread of the angular power spectrum (assumed to be a normal distribution) is specified in terms of its width W measured in radians. It is perhaps not immediately obvious that these probability diagrams do indicate some correlation between the amplitude and the bearing. To demonstrate that this is so, imagine an experiment in which the amplitude measurements are taken at some place on the ground and the bearing measurements at some other place well removed from the first. In this case there will be little or no correlation between the observa-

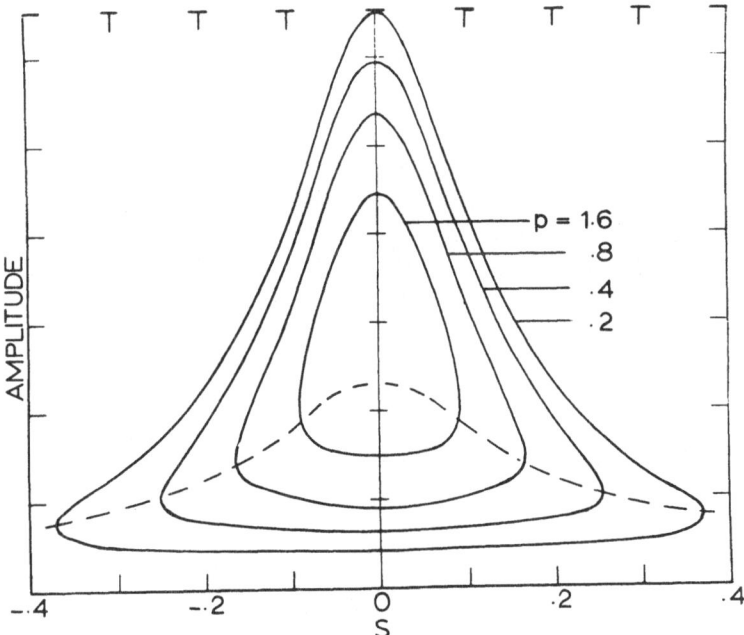

Fig. 14.3 The joint probability distribution between relative field-strength and bearing when there is no specular component. The width of the incoming fan of rays is ·1 radians.

tions at the two places and the joint probability distribution diagram will be like fig. 14.5 which is drawn for the
same angular spread of incoming waves as in fig. 14.3.

The enhanced tendency of the large bearing excursions
to be associated with low relative amplitudes when both
quantities are measured at the same place is indicated in
fig. 14.3 by the extended tails of the curves. This case
of no specular component yields the highest *correlation
ratio* (the significance of this term is discussed later in
this section) between the fluctuations in amplitudes and
fluctuations in bearings.

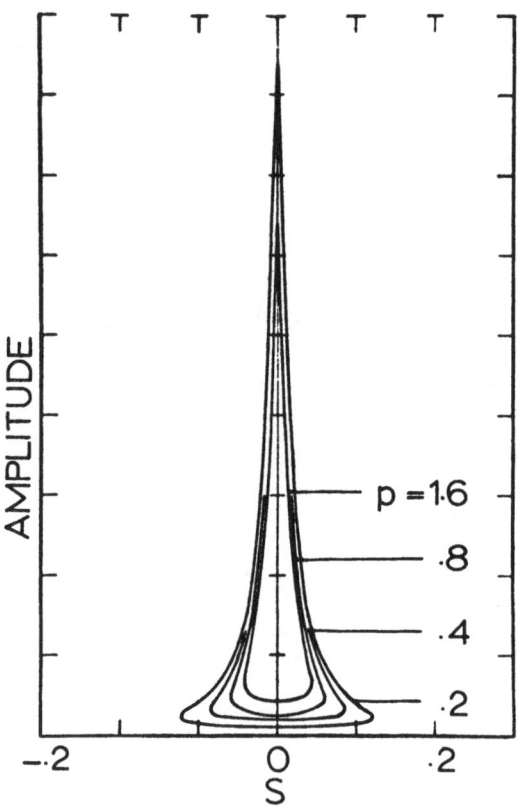

*Fig. 14.4 A similar distribution to fig 14.3 when the width
of the incoming fan of rays is reduced to ·01
radian.*

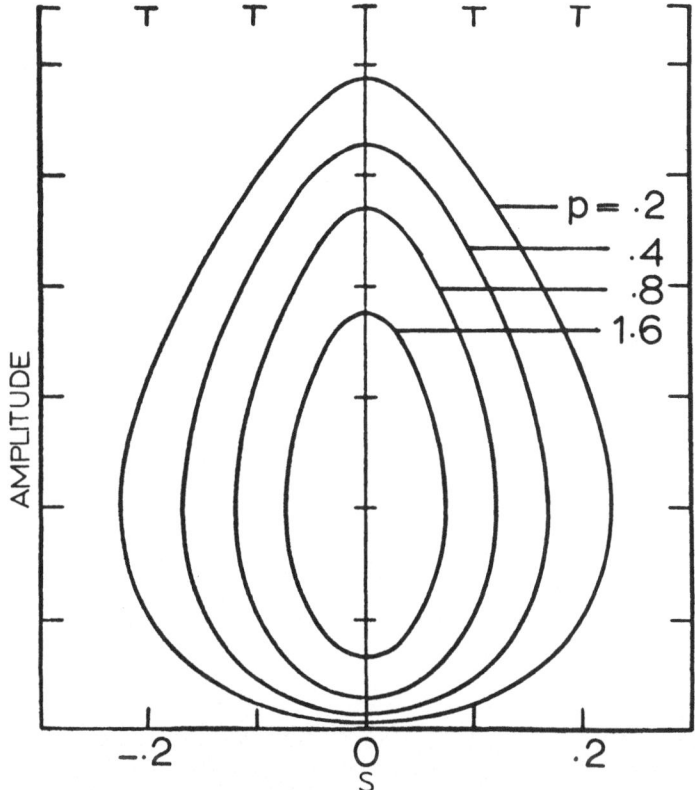

*Fig. 14.5 The distribution corresponding to fig 14.3 when the
field strength and the bearing are uncorrelated.
This could occur if each quantity were measured at
a different place.*

    The presence of a single steady specular component re-
duces the correlation because, as we have seen, the manifes-
tation of correlation in the previous case was the associa-
tion of low amplitudes with large bearing deviations and when
there is a specular component present the amplitude of the
resultant wave will seldom be very low and hence the condition which is important for correlation effects to appear
will seldom occur. As an example, the curves in fig 14.6 are
constructed for B = 8. A normalizing factor has been introduced in this case and the horizontal scale is in terms of
S/W where S = sine (wave-normal deviation) as before. It is
immediately obvious that the "tails" of the distribution have
all but disappeared in this case.

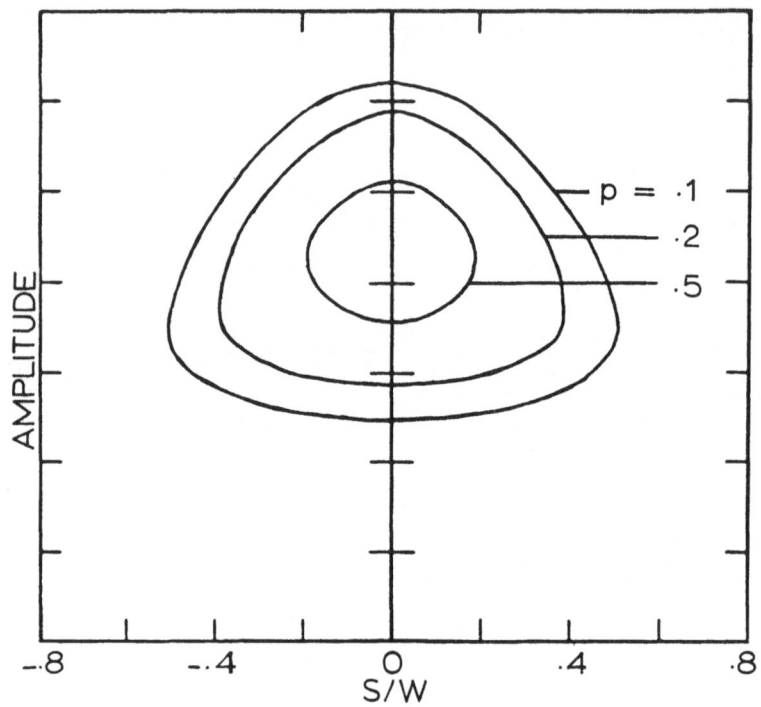

Fig. 14.6 The joint probability distribution field strength and bearing when the wave contains a specular component. In this case B (coherence ratio) = 8.

From the point of view of the physicist or engineer who is attempting to determine the structure of the *fan* of rays arriving at any particular place, these two-dimensional probability distributions are of considerable value. We have used the term "fan" here since we are, at this time, dealing only with variations in the bearing angle. Later, when considering variations in both bearing and elevation angles the incoming assembly of waves will be referred to as a *cone* of rays. It is often very difficult to decide whether observed scintillation effects arise from interference between a group of incoming waves or whether a large part of the observed directional variations arise from large-scale changes in the direction of the wave as a whole. The latter changes should be classified as day-to-day

variations of the mean direction i.e. path changes rather
than scintillations. Such path changes are not associated
with related amplitude changes so that more rounded patterns
should be obtained for the two-dimensional probability
distributions. It is possible to obtain an estimate of B
from the amplitude variations taken by themselves so that,
since B can also be obtained from the "roundness" of the
distributions, it is, at least in principle, possible to
determine the degree of real scintillation which is present.
In actual practice, if significant results are to be obtained, B should remain fairly constant. This means that periods
when there are deep amplitude fades arising from the interference of a few pseudo-specular components should be
avoided in this type of measurement.

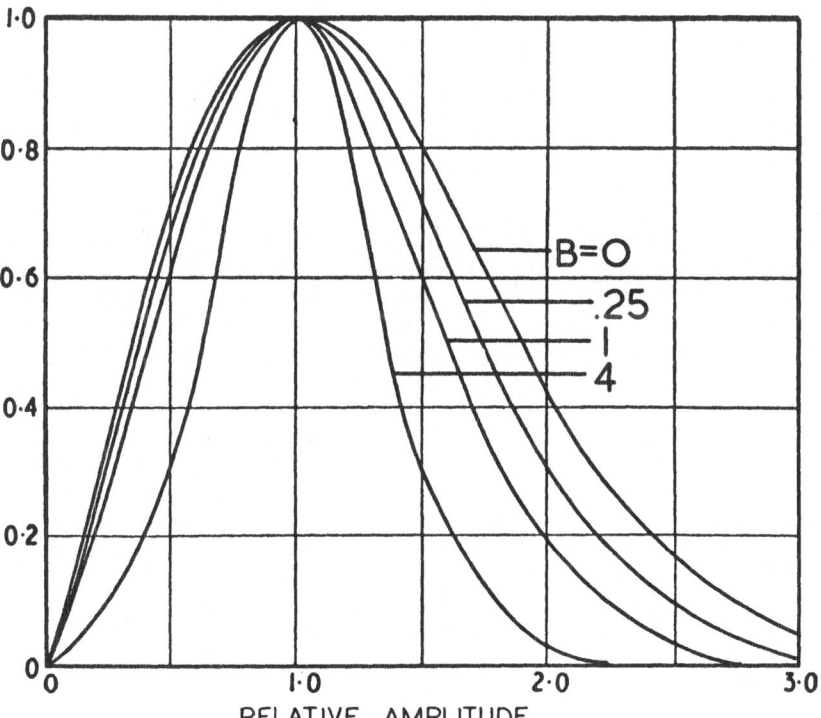

*Fig. 14.7 Amplitude probability distribution curves for de-
termining the coherence ratio B.*

For statistical observations of correlated scintillation effects to be meaningful, propagation conditions must not change during the time that the distributions are being obtained. This is generally stated as the condition that the observations must be statistically stationary. It is unfortunate that, in the case of the ionosphere, such conditions seldom hold for more than a few minutes at a time. The changes are particularly noticeable on short path transmissions where there is a large specular component. This fact is most troublesome in determining B from the amplitude distributions since the shape of the distribution does not change very much with B i.e. the shape needs to be specified with some accuracy if the results are to be useful. The way in which the amplitude distributions depend on the coherence ratio B is shown in fig. 14.7. To specify such a statistical distribution sufficiently accurately, a large number of observations and hence a long period of observation are required. In spite of this limitation, the type of result which has been obtained in this investigation of the two-dimensional distributions has indicated the general properties of the incoming fan of rays.

Further theoretical distributions can be calculated for various combinations of groups of waves with or without specular components, the groups themselves being symmetrically or asymmetrically distributed. Some of these computations are very difficult mathematically. Only one other case will be considered here since almost all others can be estimated from this and the foregoing results. This further case (depicted in fig. 14.8) is the one in which there are two main groups of waves present, each group consisting of a specular component and a symmetrical scattered component (with B = 12.5 in both cases) but with the total power in one group four times that in the other. This could correspond, for example, to the situation where a one-hop and a two-hop transmission were being received simultaneously over a relatively short path. Over medium distance paths, where many more modes may be present simultaneously, the problem degenerates into one approaching the case of no specular component. Over very long paths, it is likely that one mode or closely similar group of modes dominates all others so that, in this case, the situation is similar to a group of scattered waves with a very small specular component.

In fig. 14.8 the horizontal scale is the *ratio* of the sine of the deviation of the wave-normal direction to the

sine of the angle between the two incoming specular components. The fact that it is the sine of the various angles
which appears in these expressions is no great complication
since, for the relatively small angles with which we are
normally concerned, the sine is very nearly equal to the
angle itself (measured in radians). An experimental record
obtained by sampling simultaneous measurements of relative
amplitude and wave-normal direction which is similar to
the theoretical distribution in fig. 14.8 is shown in the
top example of results in fig. 14.9. These experimental
distributions were obtained by coupling the direction-finder
to a two-dimensional recorder so that the bearing angle
caused a pen to move horizontally while the amplitude of
the signal caused it to move vertically. The pen was
arranged to make a dot on the chart at regular intervals
(usually the interval was about one second) so that the
density of the dots is proportional to the time spent in
each element of area of the chart.

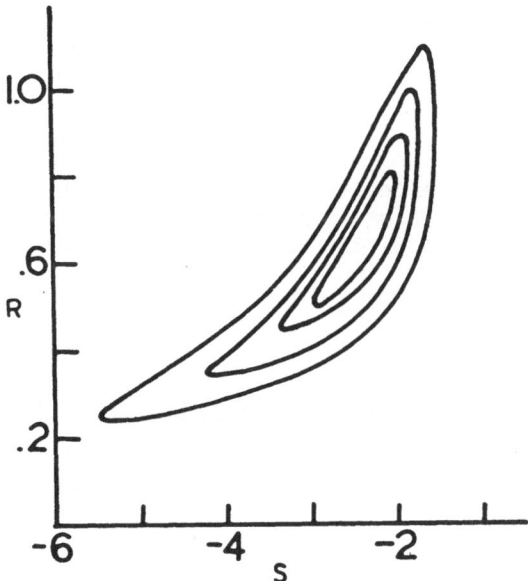

*Fig. 14.8 The joint probability distribution of relative
field strength R and wave-normal direction S for two interfering waves arriving from different directions. The variable S = sinθ/sinζ where ζ is the angle between the bearings
of the waves and θ the direction of the resultant wave-normal
direction. One wave contains 4 times the power of the other.*

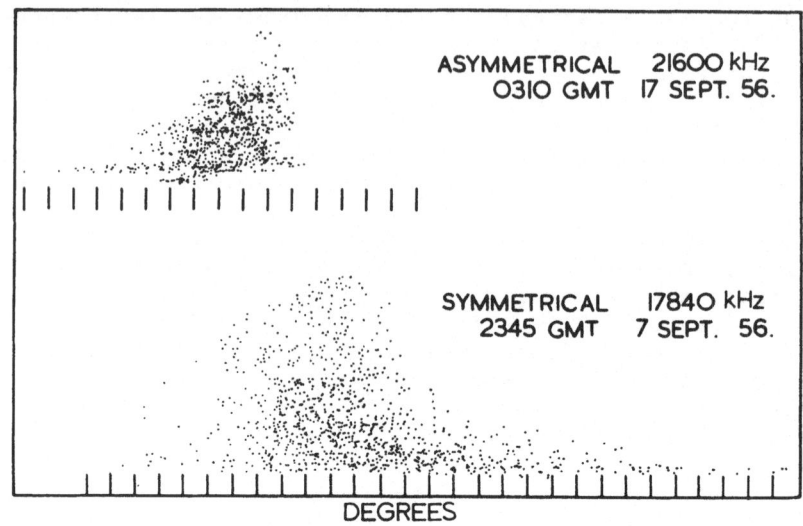

ASYMMETRICAL 21600 kHz
0310 GMT 17 SEPT. 56.

SYMMETRICAL 17840 kHz
2345 GMT 7 SEPT. 56.

DEGREES

*Fig. 14.9 Examples of experimental scatter diagrams obtain-
ed in the study of the joint probability distributions between amplitudes and bearings.*

We can apply a statistical test of the validity of the
results in this kind of investigation. It was mentioned,
without giving an exact definition of the terms, that the
correlation ratio for the two-dimensional probability distributions was highest in the first case considered (fig.
14.3) where the coherence ratio B was zero. This correlation ratio is a measure of how closely the contour lines
are clustered around the dashed line in fig. 14.3. This
line is called a *regression line* and is the line which shows
the average value of R for each small range of S. For the
case considered, B = 0, the correlation ratio η has the
value

$$\eta = \cdot 54(1 - 3\cdot 1 \ W^2) \qquad \qquad \ldots \ldots (14.3)$$

where W^2 is the variance (spread^2) of the incoming fan of
rays. Thus η has a maximum value, for very small spreads,
of $\cdot 54$. This applies, of course, only to the case considered, i.e. a symmetrical distribution when B = 0. As B increases, the value of η will decrease.

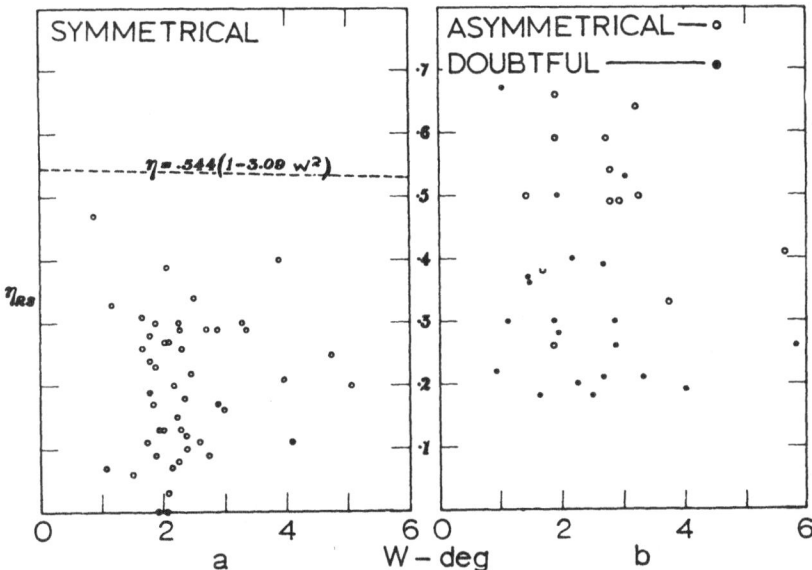

Fig. 14.10 a. The experimental correlation ratios η for
symmetrical patterns. The values never ex-
ceed the theoretical maximum of ·544.

b. Experimental correlation ratios for patterns
which may indicate the presence of more than
one major component in the incoming angular
spectrum of waves.

Experimentally derived values of η are presented in
fig. 14.10. Those in fig. 14.10a refer to the obviously
symmetrical cases and it is seen that, in fact, the values
of η always turn out to be less than ·54. For other cases,
this restriction need not hold. Especially in cases like
fig. 14.8, it is possible to obtain correlation ratios which
are larger than the value of ·54. Some such results are
shown in fig. 14.10b.

These results have been plotted with the experimentally
obtained values of W as the abscissa. There are some difficulties in relating this to the spread of the incoming fan
of rays as we will now see. If we consider the bearing
fluctuations by themselves, we find that the one-dimensional
probability distribution of the instantaneous wave-normal
direction (which we have called S) is given by

$$p(S) = \frac{W^2 dS}{2(W^2 + S^2)^{3/2}} \qquad \ldots \ldots (14.4)$$

in the case where B = 0.

If we use this expression to obtain a value for the
variance $\overline{S^2}$ of the observed fluctuations, remembering that
the extreme values which S can have (since it is the sine
of an angle) are ±1. we obtain

$$\overline{S^2} = \overline{W^2}/3 \text{ (approximately)} \qquad \ldots \ldots (14.5)$$

This is, at first sight, a reasonable result, but we
find that it depends very critically on the limits we place
on S i.e. the distribution of $p(S)$ has long tails and it is
these tails which are of most importance in determining $\overline{S^2}$.
This is unfortunate since these are the portions of the distribution which are most difficult to record accurately, ex-
pecially with the mechanical "lock-on" type of direction-
finder which has been used in the above experiments.

We can, however, obtain a fairly good idea of the way
in which the "tails" are lost in the recorded distributions.
To do this, we need to consider a quantity related to the
shape of the distribution curve, in particular a quantity
related to the peakiness or what is sometimes called the
excess of the curve. Such a quantity is the *kurtosis* K
which is given by

$$k = \frac{\overline{S^4}}{(\overline{S^2})^2} - 3 \qquad \ldots \ldots (14.6)$$

With this definition, the kurtosis of a normal or gaussian curve is found to be zero. On the other hand, the
kurtosis for a curve of the shape given by equation 14.4 is
found to be very large. This arises, of course, because the
long tails make the value of $\overline{S^4}$ in equation 14.6 very large.

If we can now obtain some experimental distributions of S under conditions where we are reasonably sure that the incoming angular power spectrum is symmetrical and of the simple form which we have postulated, we can compare the measured value of K with those predicted from the theory which includes the effects of the long tails. Now it is shown in chapter 15 that the distributions obtained for elevation angle fluctuations are essentially the same as those for the bearing angle fluctuations considered here. There is often some anisotropy in the distribution of the incoming cone of rays, i.e. the distribution plotted as a function of elevation and azimuth, usually because rays from several different propagation modes are arriving simultaneously. This yields patterns from which a plane of minimum spread can be deduced and we can be reasonably sure that, in many cases, the distribution in this particular plane will be single-humped and will thus meet the conditions assumed in the analysis. These distributions have been used in the approach to the problem described below.

The experimental distributions of S, obtained using a similar mechanical direction finder, yield values of K ranging from -1.5 to +1.5 with a mean of - .05. More than 600 distributions of the type depicted in Fig. 15.1 were analysed in this work. If we now assume that the direction-finder follows the small variations of S well but loses the large excursions i.e cuts off the tails of the distribution, we can find at what value of S the effective cut-off occurs since this must lead to experimental curves for which the kurtosis is about - .05. This value would be obtained from the theoretical curves if they were truncated at S = ± 2.2W. It then follows that, if this is a sufficiently accurate representation of the process by which the mechanical direction-finder loses the tails of the distribution, then the width in the incoming distribution W is related to the experimentally measured variance of the distribution of directions $\overline{S^2}$ by

$$\overline{S^2} = 1 \cdot 2 \; W^2 \qquad\qquad \dots\dots (14.6)$$

Even though the actual experimental distributions will tail off rather more smoothly than the truncated pattern we have assumed, it appears that equation (14.6) is a reasonable approximation to use.

Some further considerations which depend on the fact that the incoming fan of rays is made up of groups of scattered waves which have travelled different distances before arriving at the receiver are presented in chapter 18.

We have already seen that the observed spread of wave-normal directions decreases when there is a specular component present. It can be shown, as a general theorem, that whatever the shape of the distribution of energy in the incident fan of rays, the addition of a specular component (specified by the coherence ratio B) reduces the standard deviation of the wave-normal directions by the factor $\exp(-B/2)$. While the coherence ratio is a quantity which has not been intensively studied, it is of great importance in interpreting the results of scintillation measurements and some consideration of methods of measuring B is given in chapter 16.

CHAPTER 15

RELATIONS BETWEEN THE BEARING AND ELEVATION ANGLE

The three quantities of most interest in the study of scintillations are the instantaneous bearing, amplitude and elevation angle. So far this discussion on scintillation has been mainly on the relation between bearing angles and amplitudes ignoring the fact that the incoming wave can also vary in elevation angle. The foregoing thus refers to the characteristics of the projection of the actual wave normal direction on to a horizontal plane. In considering the relationship of elevation angles to amplitudes we would, similarly, be concerned with the projection of the wave-normal direction on to a vertical plane through the mean direction. In both these cases, the simple concept of a two dimensional angular power spectrum is applicable. The relationships between the amplitude and the elevation angle are similar to those obtained between the amplitude and the bearing angle (both in theory and in actual experiment) and will not be discussed in detail. The only difference is that the bearing angles in the above curves are replaced by the elevation angles which are obtained experimentally from measurements of the instantaneous ground wavelength. In the case of the relationship between the bearing and elevation angles, however, a two-dimensional picture of the angular power spectrum is required. In other words, where previously we were dealing with a specular component plus a fan of scattered rays, we are now dealing with a specular component plus a *cone* of scattered rays.

In general, it is found that the joint probability distribution is elliptical and that the probability diagram is, in the case of zero or a single specular component, a direct representation of a cross-section of the cone of incoming directions. A large number of experimental results have been obtained for this type of joint distribution. A few samples are shown in fig. 15.1 in which diagrams the bearing is plotted horizontally and the elevation angle vertically.

These results were obtained with a further type of direction-finder specially developed for the purpose. It was based on the equipment used for obtaining the continuous records of the fluctuations in bearing direction but was

also able to provide continuous records of the fluctuations
in the elevation angle i.e. it provided both the horizontal
and the vertical components of the fluctuations in the
incoming wave-normal direction.

A very marked thinning of the ellipse is expected when
there are two specular components present. For example, if
each specular component was considered to be fading in amplitude but not accompanied by any scattered cone, the ellipse
would degenerate into a straight line. Apart from their
indication of the presence of multiple specular components,
the ellipses may indicate some anistropy in the irregularities in the ionosphere. However, since there have generally
been a large number of encounters with the ionosphere, it
could be difficult to predict what the shape would be in any
particular case. One purpose of this type of investigation
in the study of long-distance propagation is that it indicates the beam width and shape of antenna polar diagram
which is necessary in order that most of the available energy
be collected.

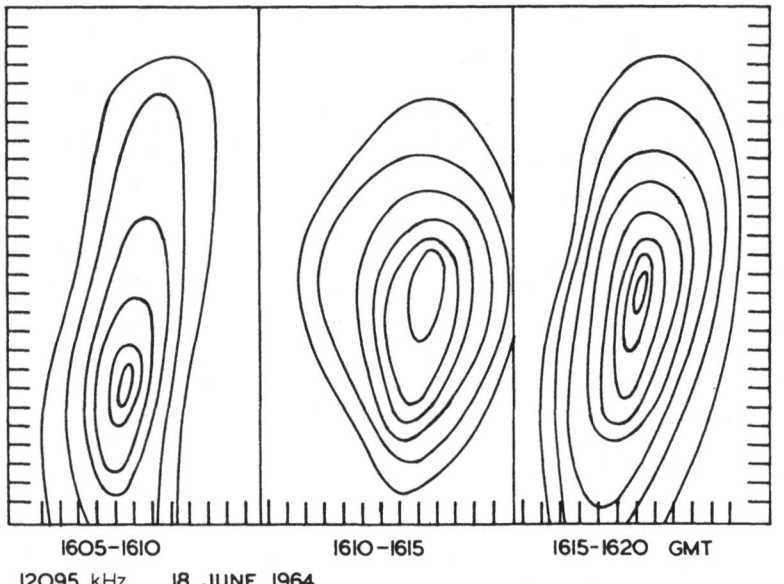

1605-1610   1610-1615   1615-1620 GMT

12095 kHz   18 JUNE 1964

Fig. 15.1 Experimental joint probability distributions of
           bearing angle (horizontally) and elevation angle
           (vertically). The diagrams are to scale with de-
           grees indicated by the tick marks.

It has already been pointed out that the presence of a specular component in the incoming cone of rays reduces the spread of the recorded distribution of wave-normal directions. This reduction is effective for both the elevation angles and the bearing angles so that the shape of the two-dimensional distribution is unchanged and the value of the coherence ratio often need not be known explicitly.

The interpretation of the spreads in the vertical direction is, at first sight, rather more complicated than in the previous case of the bearing angles. It will be remembered that we are concerned with measurements of the instantaneous direction of the wave-normal to the corrugated or wrinkled wave-front which has developed as a result of the interference between the component rays in the incoming group of rays. In practice, these corrugations are rather more complex than those which would arise from the interference effects between the incoming rays by themselves because there is also a group of similar waves arriving at the receiving antennas after having been reflected from the ground. These reflected waves will generally have no effect on the measurements of the bearing angle of the wave-normal but may affect the measured vertical angle since this angle is not obtained directly but is deduced from measurements of the wavelength along the ground as shown in figure 4.1a. We can consider the situation in a similar way to the approach used in obtaining figure 3.2. It is found that, if there is only one downcoming ray, then the measured ground wavelength yields the correct value of the vertical angle of incidence whatever the ground reflection coefficient and whatever the type of antenna used i.e. irrespective of whether the antennas receive horizontally or vertically polarized waves (although both must receive the *same* polarization). When there are two incident rays, it is found that the resultant effect may be represented in a form which is the same as if there were no ground present, *provided that* each ray is modified in amplitude by an amount which is equivalent to the vertical directivity curve of one of the antennas with its ground reflection. This means that the effect of the ground is relatively unimportant when vertically polarized waves are being received (since then the directivity curve does not change very rapidly with the incoming elevation angle) but can be very important when horizontally polarized waves are being received. In the latter case the directivity curve is strongly dependent on the elevation angle. Even in this case, if the vertical spread of the waves is not too

large and the mean elevation angle not too small the indicated
vertical spread will not be affected very much by the presence
of the ground. This is the case on very many occasions in the
study of very long distance transmissions where the mean
elevation angle is generally of the order of 15 - 20°. The
main effect of the ground reflections in measurements using
horizontally polarized waves is the introduction of an asy-
metry into the downcoming angular power spectrum so that
the apparent centre (or mean direction) of the waves is
shifted upwards.

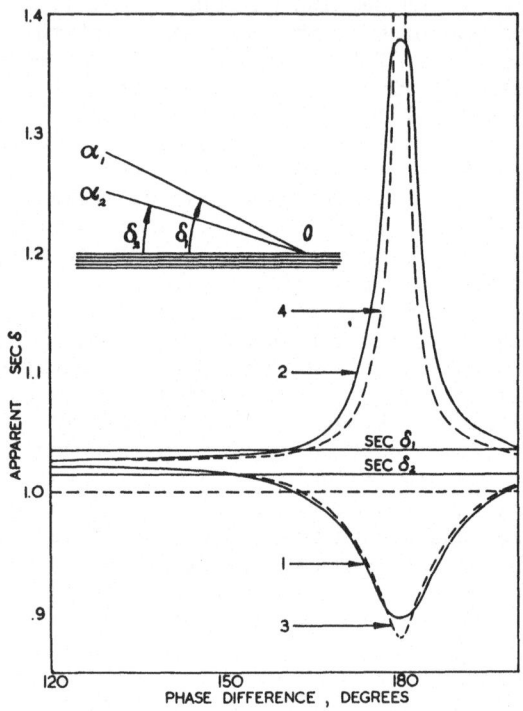

Fig. 15.2   The way in which the indicated wavelength λ_g along
the ground can vary when there are two interfering
waves present. The apparent value of $\sec\delta = \lambda_g/\lambda$
is shown. For this diagram $\delta_1 = 15°$, $\delta_2 = 10°$

Curve 1 $\alpha_1/\alpha_2 = \cdot 9$ vertical polarization
  "   2 $\alpha_1/\alpha_2 = 1/\cdot 9$
  "   3 $\alpha_1/\alpha_2 = \cdot 6$ horizontal polarization
  "   4 $\alpha_1/\alpha_2 = \cdot 7$

An important interference effect which arises when there are two or more waves in the incident angular spectrum is that the instantaneous ground wavelength can be *less than* the free-space wavelength. This means that the indicated value of the elevation angle δ can not be found on these occasions (since sec δ < 1) so that it is necessary for any statistical analysis to be carried out initially in terms of the indicated ground wavelength rather than the elevation angle. The resulting quantities (variance, for example) can however, be expressed in terms of real angles. Some examples of this behaviour are shown in figure 15.2. The two incident rays of amplitude α_1 and α_2 (shown in the inset) arrive at elevation angles of 15° and 10° respectively. As the phase difference (measured at some arbitrary point, in this case at O) between the two waves changes the indicated ground wavelength also changes. In this diagram the ground wavelength is expressed as its ratio to the free space wavelength, and this ratio is, of course, sec δ (where δ is the indicated elevation angle). Four cases are considered. Firstly, a ground reflection coefficient of +1 (corresponding to vertically polarized waves) is assumed with α_1/α_2 = .9 and 1/.9 yielding the full-line curves 1 and 2. Secondly, a ground reflection coefficient of −1 (corresponding to horizontally polarized waves) with α_1/α_2 = .6 and .7 gives the dashed curves 3 and 4. In both types of polarization large excursions of the sec δ curve into both the upper and lower parts of the graph are observed when the phase difference at O approaches 180°.

If the distribution of power in the incoming two-dimensional angular power spectrum is simple i.e. there are no multiple humps or large asymmetries in any section through the distribution then the recorded two-dimensional probability pattern is a scale diagram of the incoming distribution; some sample probability patterns are shown in fig. 15.1. It will be recalled that, because these were obtained using a mechanical direction-finder, the tails of the distributions will not appear in these patterns. It will also be recalled that, if the specular component can not be assumed to be zero, it must be measured before the actual values of the spreads of the incoming cone of rays can be obtained.

The patterns in fig. 15.1 are drawn to scale in the sense that the vertical scale of elevation angle and the horizontal scale of bearing angle are both the same. The

question arises as to whether the obvious lengthening of the
elliptical patterns in the vertical direction arises from a
single simple but spread-out distribution or from the super-
position of several distributions representing different
propagation modes and thus centred on different elevation
angles. This question can be resolved by again finding the
kurtosis of the distributions along and at right-angles to
the major axis of the ellipse which best fits the pattern.
It can be shown that, if the patterns arise from a superpositioning of two independent patterns, there will be a
relationship between the measured kurtosis and the measured
variance taken in a direction parallel to the line joining
the centres of the two independent distributions. From such
an analysis it can be deduced that the major part of the
vertical spreading in the case of near-antipodal propagation
does, in fact, arise from such a superpositioning of two or
more distinct groups of waves.

It has been pointed out in the above that the coherence
ratio B is of considerable importance in determining the
type of distribution which is to be expected in any particular case. Since this ratio is of such importance, some
discussion of the methods of measuring it will now be given.

CHAPTER 16

THE COHERENCE RATIO

Some importance can be attached to a knowledge of the factor B. Although it is the value of B which largely determines the characteristics of the scintillations at any place, it is not easily obtained from measurements of these phenomena. As an example of this, consider the fluctuations of the amplitude of a received wave. The expected statistical distribution can be obtained either by a direct calculation *ab initio* or from fig. 14.3. In this figure the relative amplitude is plotted vertically and the bearings horizontally. If only the amplitude distribution is required, it is obtained by summing the probabilities in strips horizontally, each strip being a small increment of amplitude in height. This yields the so-called *marginal distribution* which, in this case, is for the amplitude only. If this process is performed for various values of B, a set of curves is obtained as in fig. 14.7. In practice, the mean amplitude of the signal must be found from the measured values i.e. any information from experimental amplitude probability curves must be derived from the *shape* of the curves. The curves in fig. 14.7 have all been plotted with the same value of the maximum relative probability so that the shapes can be compared. It is immediately obvious that, even under favourable conditions when the fading is statistically stationary (i.e the statistics do not change over the period of the measurement), the accurate determination of B from experimental curves of this type is difficult enough when B is large but it is next to impossible when B is small.

In point of fact, however, the mean value of the amplitude (obtained by averaging out the scintillations) can and usually does change considerably over times of the order of five minutes. If the scintillation fading rate is slow, it may take much more than five minutes to obtain enough experimental points to allow the the probability distribution to be obtained even approximately. To obtain sufficient points to enable a distinction to be made between the curves for B = .5 and B = .7 similar to those in fig. 14.7 could well take several hours.

One method of overcoming this problem is to take a
large number of observations simultaneously at well
separated receiving points so that there is virtually no
correlation between the fading at the various receivers.
The time required to obtain a given amount of data is then,
of course, inversely proportional to the number of receiving
points employed. The logical extension of this is to extend
the system to such a very larege number of receiving points
that only one measurement at each point gives enough information. We are then substituting the variation of the signal
strength with position for its variation in time, the two
approaches being equivalent. This latter method leads to a
complicated installation, in practice requiring an extensive
array of antennas, feed-lines and an accurately calibrated
switching system. It is essential that no amplitude differences additional to those arising from the fading itself
be introduced anywhere in the system, or, if they are, that

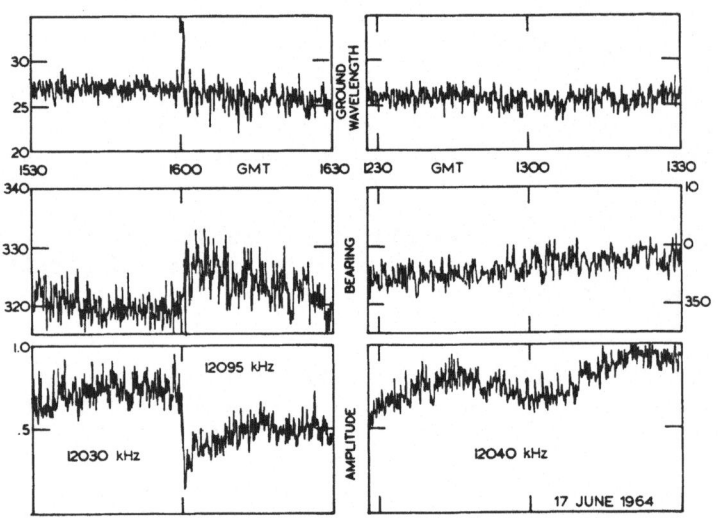

*Fig. 16.1 Experimental records of ground wavelength (elevation
angle), bearing and amplitude of three different stations. The
character of the variations in the first two quantities tends
to change less than does that of the amplitude.*

a series of measurements in the time domain be made at each
antenna simultaneously so that any systematic amplitude
changes can be removed from the results by a time-averaging
process. This is, of course, essentially a calibration of
the system.

Many of the slow changes in amplitude which are so
troublesome arise from relatively slow changes in absorption
along the path so that both the specular and the scattered
components are affected similarly. In this case, the value
of B does, in fact, stay constant for rather longer periods
of time, since B is a ratio and does not depend on the
absolute value of the signal strength. While such a change
in mean amplitude over the recording period rules out any
method of measuring B which depends on obtaining statistical
distributions of amplitude, it does not affedt a method
which depends on directional observations of the wave-front
for example.

As an example of the differences in the characteristics
of scintillation fading between the angle fluctuations and
the amplitude fluctuations we can refer to fig. 16.1. Here
the continuous records of the elevation angle (actually the
ground wavelength), the bearing angle and the relative
amplitude of the signal are shown. These are all taken
simultaneously and were obtained using the same direction-
finder as was used to obtain the results discussed in
chapter 15. Although the mean level of the amplitude fluctuations has changed considerably during the period of the
record the characteristics of the angular fluctuations persist indicating that the only effect occurring was a slow
change in the overall power of the whole angular spectrum
of waves making up the incoming wave system. It will be shown
in the following that measurements of the phase difference
distribution between spaced antennas can lead to quite
accurate estimates of the coherence ratio.

We are thus led to a consideration of what properties
of the received wave can be determined using a system of
spaced antennas and it is at this stage that it becomes convenient to think in terms of the generalized auto-correlation
function rather than the angular power spectrum.

We have stated that the generalized auto-correlation function is the same across the ground as it is across the wave-field just as it leaves the ionosphere. The relation between the generalized auto-correlation function $\rho(x)$ and the correlation function $r(x)$ for the phase deviations introduced by the ionosphere was stated previously to be

$$\rho(x) = \exp - \overline{\phi^2}(1 - r(x)) \qquad \ldots\ldots (16.1)$$

where $\overline{\phi^2}$ is the mean square phase deviation in the screen (ionosphere).

In the case of irregular (or random) phase deviations which is being considered here, $r(x)$ will tend to zero as x becomes large since the phase variations at large separations in the ionosphere will arise from different irregularities and will hence be uncorrelated i.e. $r(\infty) \to 0$. At zero separation, the correlation between two points is perfect i.e. $r(0) = 1$. Since $r(x)$ may often be a function which decreases monotonically from 1 to 0 as x increases, $\rho(x)$ will have the form drawn in figure 16.2. It will be noticed that, at large separations $\rho(x) = \exp - \overline{\phi^2}$. We mentioned previously that

$$B = \frac{\exp(- \overline{\phi^2})}{1 - \exp(- \overline{\phi^2})} \qquad \ldots\ldots (16.2)$$

and hence $\exp(- \overline{\phi^2}) = \frac{B}{1 + B}$. Now $r(x)$ is a function directly dependent on the phase fluctuations in the ionosphere. It can be measured directly only in a plane immediately below the ionosphere where only the phase fluctuations are important and the amplitude fluctuations have not yet had time to arise from the interference effects. On the other hand, $\rho(x)$ can be obtained from measurements taken anywhere in the wave-field; for example, it may be obtained from measurements taken along the ground. In order to see how $E(x)$ may be evaluated from measurements which can be made in practice, it is necessary that we look at it a little more closely. Now

$$\rho(x) = \Big\langle E(X) \cdot E^*(X + x) \Big\rangle$$

where $E(X)$ is the wavefield at any point X and $E^*(X + x)$ is the complex conjugate of the wavefield at a point $X + x$. The diagonal brackets indicate that the product is to be averaged over all values of X.

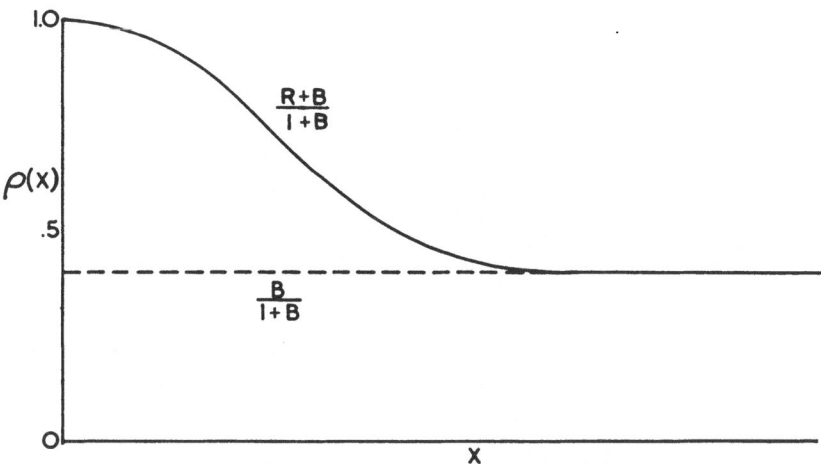

Fig. 16.2   The generalized autocorrelation function $\rho(x)$, where x is the separation between two antennas. If the coherence ratio is B, $\rho(x)$ attains the value $\frac{B}{1+B}$ at large separation distances.

Since we require that $\rho(0) = 1$, it is necessary either to apply a constant multiplying factor to all the E's or to divide the averaged product by $\langle E(X) \; E^*(X) \rangle$. It is slightly simpler to adopt the first procedure. Now a wavefield E may be represented by an amplitude term and a phase term. Under the relatively minor restriction that the scattered components are symmetrically distributed about the specular component, it is then found that

$$\rho(x) = \langle E(X)E(X+x) \; . \; \cos(\phi(X)-\phi(X+x)) \rangle \;\; .. \;\; (16.3)$$

where $\phi(X)$ is the phase of the wave-field at any point X and thus $\phi(X+x)$ is the phase at the point X+x. Thus the generalized auto-correlation function can be expressed in terms of the average of the product of two quantities, one of which involves the amplitudes of the field at two points and the other involves the phase difference between the signals observed at two points. These are the two basic types of measurement which can be carried out using two separated antennas. It should be noted that these two quantities (the product of the amplitudes and the cosine of the phase difference) are not independent. There is the same type of correl-

ation between them as has already been found to exist between
the amplitude and the wave-normal direction in the measurements using a direction-finder which were considered previously in chapter 14. It is only recently that a solution has
been found to the mathematical problem of specifying the way
in which these two quantities (and other similar quantities)
vary when the ratio B has a wide range of values. Approxi-
mations valid when B is either very large or very small have,
however, been known for some years. The complete expressions
are fairly complicated but the results may be presented in
the form of probability curves.

It has been noted previously that there are difficulties
associated with the use of amplitudes in obtaining the statistical distributions which are needed in this type of work.
We will therefore concern ourselves mainly with phase difference distributions in which case the measurements which are
made will provide some quantity which is related to the cosine term in the expression for $\rho(x)$ which has been given in
equation 16.4. It is found that, in fact, no measurements on
the amplitude distributions are necessary since the generalized auto-correlation coefficient may be written in the form

$$\rho(x) = \frac{R + B}{1 + B} \qquad \ldots\ldots (16.4)$$

and both B and R (discussed below) can be obtained from
measurements of phase difference distributions.

In this expression we have introduced a new parameter
R which is different from $r(x)$ which was used previously.
Now we know that a screen in which there are only phase
fluctuations (which are described by $r(x)$) gives rise to a
specular component and a range of scattered components. We
also know that, if the fluctuations in the screen had been
only amplitude fluctuations with a mean value of zero (i.e.
there were both positive and negative amplitude fluctuations,
the latter representing waves with a 180° phase change) the
screen would give rise to scattered waves only with no
specular component. We can postulate that there exists some
pure amplitude screen which would give rise to the same
angular power spectrum of scattered waves as the scattered
part obtained with the actual phase screen we are considering.

We can describe this fictitious amplitude screen by the
function R which is its autocorrelation function. Of course
R varies with the distance x similarly to the way in which
r(x) varies but it is not the same function. In order to
find $\rho(x)$ from the expression 16.4 we need to know both R
and B. Fundamentally, this indicates that two independent
measurements are required. Now R (like r(x)) tends to zero
as x becomes very large so that if we take two widely separated antennas we will approach the condition that

$$\rho(x) \rightarrow \frac{B}{1 + B} \text{ as } x \rightarrow \infty$$

We find that, in this case, the phase difference distribution between the two spaced antennas depends only on
the parameter B i.e. B can be obtained from phase difference measurements made on two widely spaced antennas. The
basic measurement consists of obtaining the probability
distribution of the phase difference χ between the two
spaced antennas. From this some *descriptive parameters*
such as the standard deviation of the spread or the average
value of $\cos\chi$ is obtained. There is then a unique value of
B which corresponds to this observed value.

If the whole shape of the generalized auto-correlation
curve is required, a series of measurements using different
antenna spacings must be made. For each of these spacings
the appropriate descriptive parameter is obtained. Since B
is already known, the value of R can be found from computed
curves relating the parameter to R for each value of B.
Such a set of curves using the standard deviation of the
spread as the descriptive parameter is shown in fig. 16.3.
Then $\rho(x)$ is found from equation 16.4.

The main disadvantage of the above approach arises from
the fact that B is obtained from measurements made with two
antennas spaced so far apart that R has decreased to a negligible value; the separation required for this condition to
be fulfilled may be quite unreasonable in cases where the
angular power spectrum is very narrow. Also, unless it is
known *a priori* that a given separation is adequate some
criterion, such as the fact that the measured distributions
are the same at two different large separations, is required
to indicate when an adequate separation has been achieved.

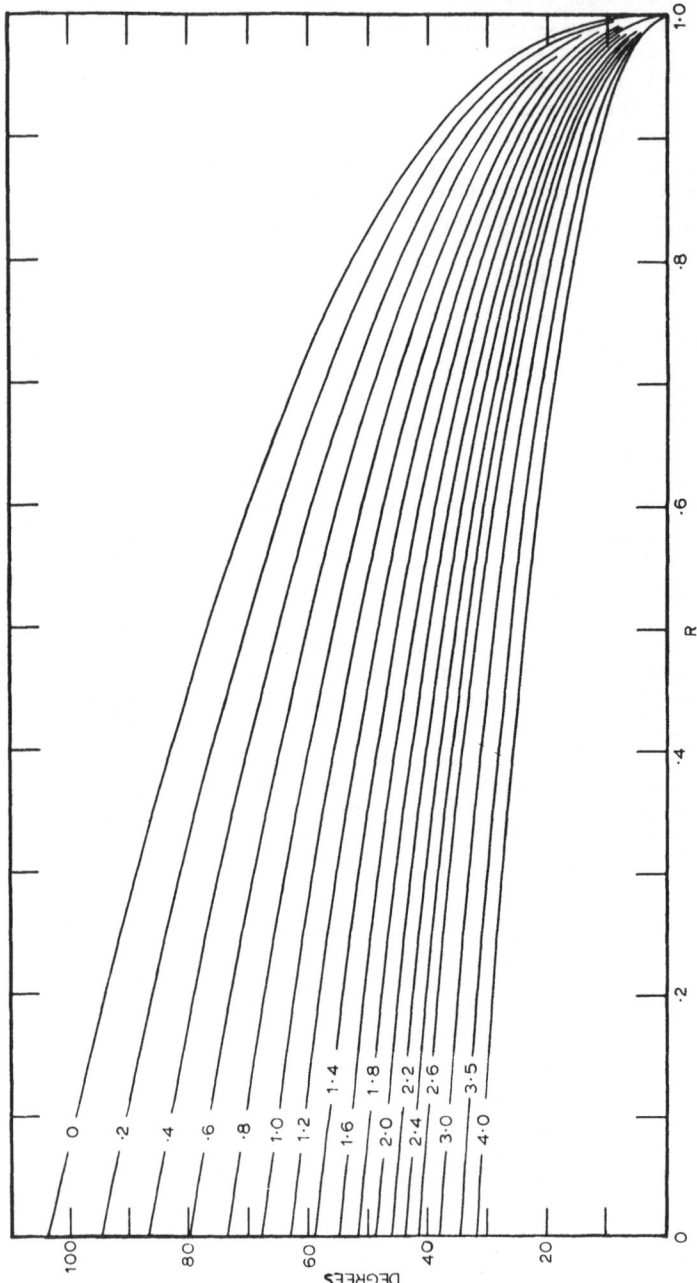

Fig. 16.3 The standard deviation of the phase difference between two antennas as a function of R and B. The coherence ratio B is given by the parameters on the curves.

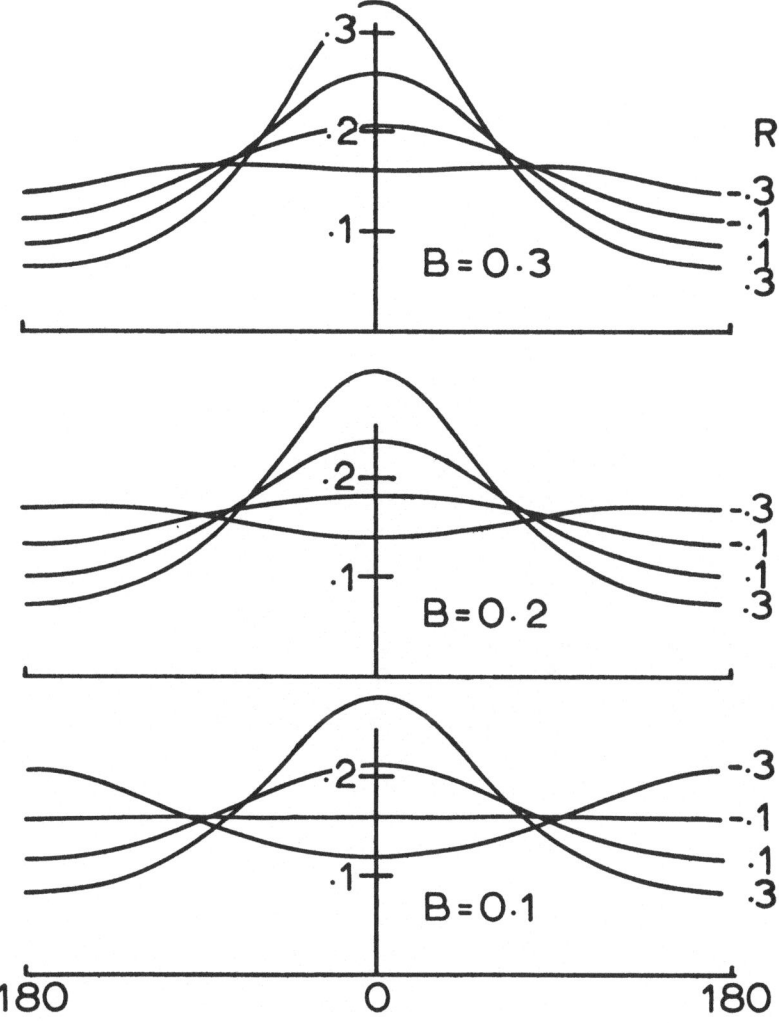

Fig. 16.4. *Typical phase-difference probability distribution*
curves for selected values of B and R.   The
position at which the maximum occurs may change
by 180° when R is negative.

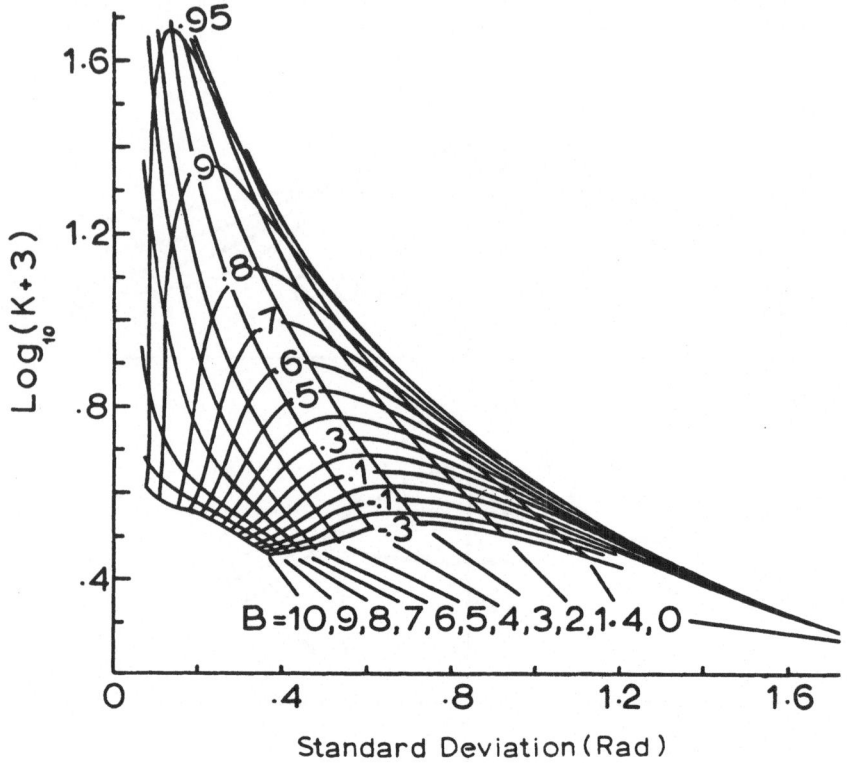

Fig. 16.5.   *Curves for obtaining the correlation function R*
and the coherence ratio B from measured values
of the standard deviation and the kurtosis K.

As has been mentioned previously, $\rho(x)$ is a quantity
which is directly related to the irregularities in the ionosphere for the case where the wave has made one transit
through the ionosphere. In the case of long-distance propagation, it is not easily related to ionospheric conditions
but it will give the angular power spectrum of the incoming
fan of rays or, if the antennas are spread out in two
directions on the ground so that $\rho(x,y)$ is obtained, it will
give the angular power spectrum of the incoming cone of rays.

The importance of this method lies in the fact that the
actual shape and not merely the width of the angular power
spectrum can be obtained.

An alternative approach, which depends for its success on
the fact that measured phase-difference distributions can
be specified with some accuracy, is possible. The phase
difference curves for different combinations of B and R
have different widths and different shapes so that they
can be identified if suitable descriptive parameters are
chosen. One pair of such parameters is the standard deviation σ and the kurtosis K of the curves. A set of typical
phase difference probability distribution curves is shown
in fig. 16.4. These have been computed for four values
of R (R = .3, .1, -.1, -.3) for each of three values of B
(B = .1, .2, .3). The negative values of R have been
included since, although we have, in previous work, postulated for simplicity that R decreases monotonically from
1 to 0 as the separation increases, there is no compelling
reason why this should occur. All that is certain is that
R = 1 when the separation is zero and that R lies between
± 1 at separations greater than this. If the irregularities
in the screen are random, then R will tend to 0 as the separation becomes very large. In naturally occurring phenomena
where inhomogeneities are present in a medium which can
support wave-motions it is common for a compression (for
example) to be succeeded by a rarefaction so that there is
a tendency for the autocorrelation function describing these
inhomogeneities to become somewhat negative before it settles
down to its final value of zero at large spacings.

The relationship between the σ and K of these curves
and the B and R which also defines them is shown as a series
of curves in fig. 16.5. It is seen that the curves become
very closely spaced for small values of B so that the accuracy is not high in this region. A better approach has been
found to be to fit the experimentally determined phase
difference distributions to the theoretical distributions by
a least mean square technique; a series of such experimental
distributions and the fitted curves are shown in fig. 16.6.
The actual experimental points are shown only when they
differ from the theoretical curves.

The task of determining the distribution of the phase
difference between the two antennas can require quite a
considerable amount of complicated processing and recording
equipment. It is worth mentioning here that the whole
process can be carried out quite easily using the rotating
interferometer. We will not discuss this technique in detail
beyond pointing out that it involves determining the densities of the pen marks at the points a, b, c, ... in fig.
4.6 and the slopes of the crossover points a', b', c', ...
By the densities is meant the ratio of the number of lines
marked to the number left blank. If the observing time is
such that 10 rotations of the interferometer are made, then
the density can be specified in units of .1 up to .9 (one
line only left blank). This range of densities enables
values of B up to 1.5 to be measured. For greater values
of B the method is indeterminate but the result can then be
obtained from the rate at which the average crossover from
"pen marking" to "pens not marking" occurs. The required
curves for carrying out this type of analysis are given in
WHALE and BANNISTER (1966). In the first case, i.e. where
the line density is measured, the formula to be employed
is particularly simple. For, if the probability that the
pen will mark is given by P, we have the result

$$P = 1 - \frac{1 - R}{2} \cdot \exp - B.$$

When one antenna of the rotating interferometer is used
in conjunction with a fixed *remote* antenna so that R = 0,
this relation becomes

$$P = 1 - \frac{\exp - B}{2}.$$

This method is by far the most convenient known for
measuring small values of B.

The other case where the rate of change of the edges of
the patterns are used can not be expressed as simply as this
since some rather complicated functions appear in the analysis. In this case it is, however, quite simple to use the
curves which have been computed (*loc. cit.*).

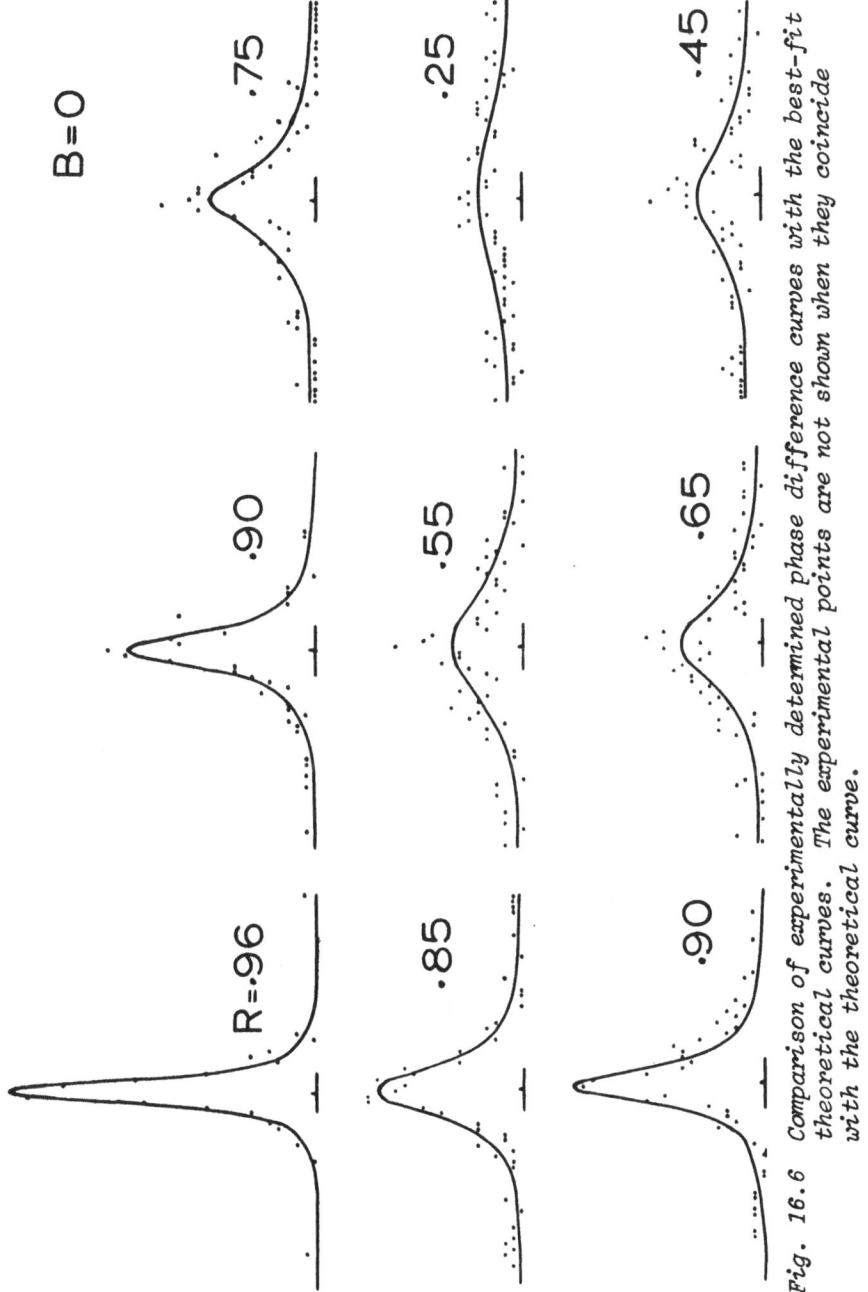

Fig. 16.6 Comparison of experimentally determined phase difference curves with the best-fit theoretical curves. The experimental points are not shown when they coincide with the theoretical curve.

The generalized autocorrelation function falls off as
the distance between the two observing points increases, but
there is no reason why it should fall off in the same way
when the line joining these observing points has different
orientations; in other words, the shape of an experimental
curve (corresponding to the idealized curve in fig. 16.2)
could be quite different when the antenna separation occurred
along the ground in a NS line, say, from that obtained along
an EW line. In many investigations, especially those concerned with the determination of ionospheric winds (i.e. the
way in which irregularities move in the ionosphere) it is assumed that the shape of the two-dimensional generalized auto-
correlation function measured on the ground is approximately
elliptical. This means that, although the widths of the
peaks of the curves may be different along the two directions
postulated above, there is a smooth transition from one curve
to the other as the direction along which the antennas are
separated is changed from NS through NE-SW (say) to EW. Some
measurements which throw considerable light on whether this
smooth transition actually occurs have been made by B.W. Hicks
using 100 antennas set up in a 10 by 10 array with 10 meters
separation between adjacent antennas. The measurements were
made using amplitude fluctuations but, since so many antennas
were employed, the usual problems of lack of statistical
stationarity did not arise. Some typical results are shown
in figs. 16.7--16.9 where the numbers on the curves are actually the correlation coefficient of the amplitude fluctuations;
this quantity is closely related to the factor R which appears
in expression 16.4 for the generalized autocorrelation function.

The contours in fig. 16.7 are almost elliptical and show
the simplest type of pattern which could be obtained. Those
in fig. 16.8 are almost elliptical but the orientation of the
inner ellipses is different from that of the outer ellipses.
The bearing of the station is along the dotted lines in these
two figures. The inner ellipses in fig. 16.8 probably arise
from the diffraction effects of small irregularities in the
ionosphere while the outer ellipses arise from the interference between the various rays which are almost specular and
make up the total pseudo-specular component.

A much more complicated pattern is shown in fig.16.9
which was obtained on a long-distance transmission on 15270
kHz. The rapid fall-off of the correlation function indicates
the presence of a considerable amount of widely scattered
radiation in this case.

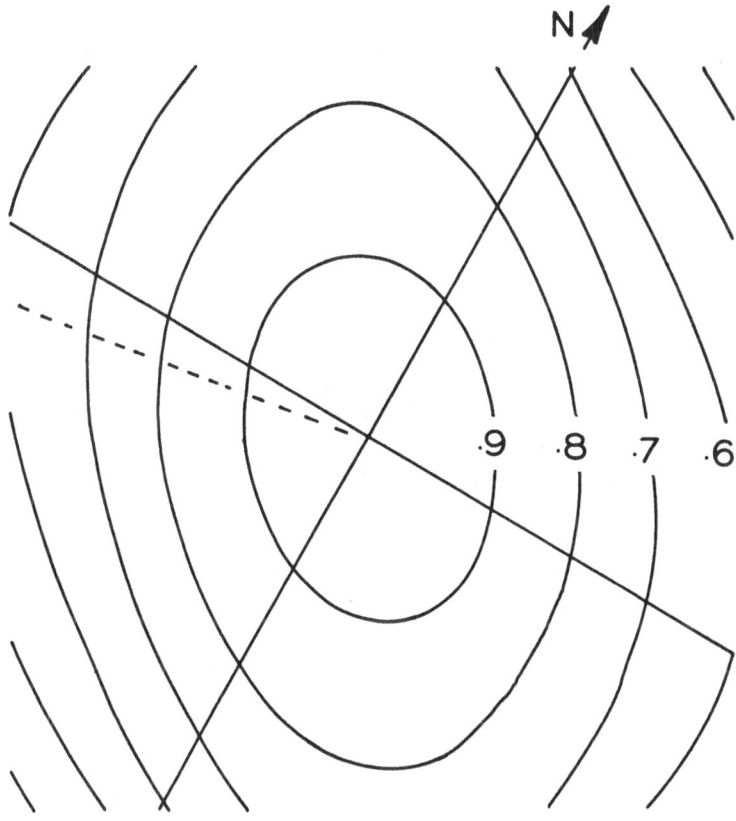

Fig. 16.7 Amplitude correlation coefficients for trans-
mission from VMG (Australia) on 12005 kHz at
0112 GMT on 24 October, 1966.

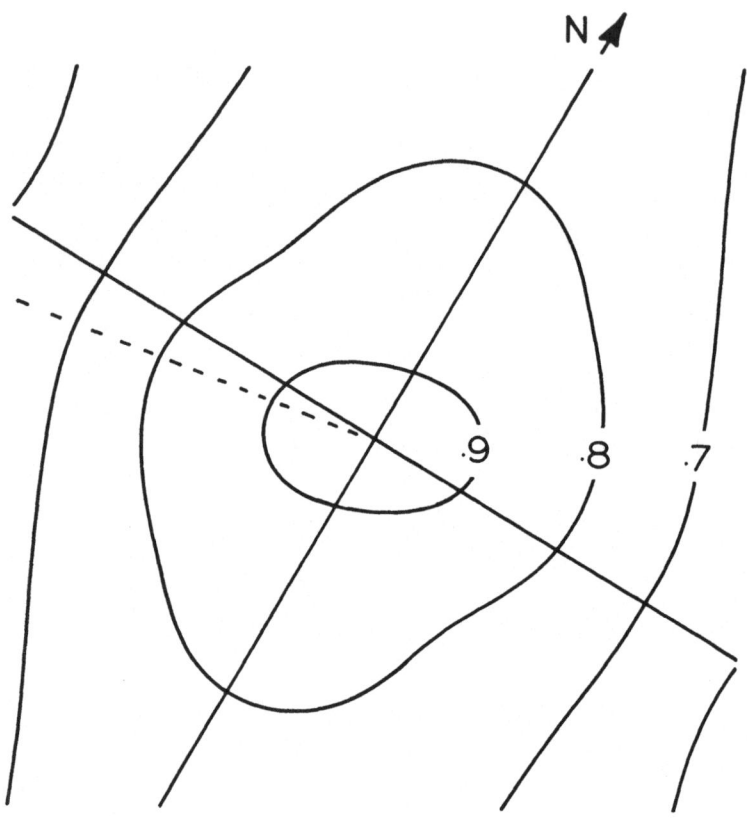

Fig. 16.8 Amplitude correlation coefficients for trans-
           mission from VMG (Australia) on 12005 kHz at
           0352 GMT on 24 October, 1966.

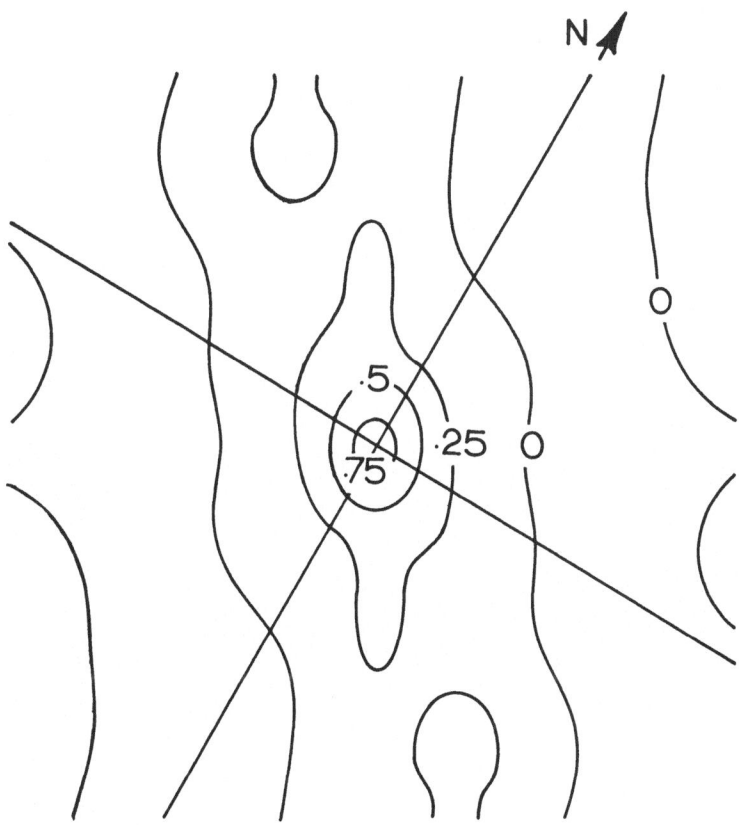

Fig. 16.9 Amplitude correlation coefficients for transmission of U.S. Armed Forces program on 15270 kHz at 0320 GMT on 27 March, 1967.

CHAPTER 17

THE ANTIPODAL AREA WHEN THE IONOSPHERE IS SMOOTH

Although most of the work so far has been concerned
with fluctuations in the direction of arrival of radio waves
it has been pointed out that these are often associated with
changes of the amplitude of the signals i.e. with the signal
strength. In analogy to this effect, the angular spreading
of waves as they propagate round the earth leads to effects
of the strength of the received waves. For example, in the
study of day-to-day bearing deviations it was seen that there
were, in general, particular characteristics associated with
the point which was exactly antipodal to the transmitter,
the *antipodal point* being, by definition, the point through
which all great circles from the transmitter pass. In the
radio propagation case, the actual existence of an antipodal
point in practice depends on several generally unrealized
idealizations, since it exists only if all rays leaving a
transmitter have, as their traces on the earth's surface,
great circles. This condition implies that the earth's
surface is smooth and spherical and that the ionospheric
reflecting surface is smooth, spherical and concentric with
the earth. Not one of these conditions is met completely
in practice; all that can be stated with confidence is that
the earth's surface is nearly spherical. In spite of this,
let us consider the power distribution over the ground in
the vicinity of the point which is geometrically antipodal
to the intersection which a straight line through the
transmitter. The *geometrical antipodal point* is defined
as the intersection which a straight line through the transmitter and the centre of the earth makes with the surface
on the opposite side of the earth. This definition is
adopted since it does not depend on the form of any ray
paths i.e. it does not refer to great circle paths.

Under the specified idealized conditions given above,
every ray from T in fig. 5.1 (except the limiting vertical
ray at T) will, after a number of reflections, pass through
the vertical line at AP. Close to the point AP, therefore,
the energy contained between the earth and the ionosphere per
unit area of surface (i.e in each vertical tube of unit
cross-sectional area) will decrease with distance from the
point AP. If the angular distance (i.e the distance measured as the angle subtended at the centre of the earth) from

AP is called μ the energy per unit vertical column is given
by

$$P(\mu) \propto 1/\mu$$

The variation of power density along the surface of
the earth is of more practical interest. The difference
between this case and the former is that, now, only rays
with certain particular elevation angles can reach a particular point on the earth's surface. The power distribution
will, on the whole, be essentially the same as above because
each complete group of rays which leaves the transmitter
at a particular elevation angle will illuminate a ring
around the antipodal point. A slight change in the
elevation angle at the transmitter can cause the ring
either to contract into a point at the antipodal point
or to spread out into a larger ring of radius μ, say.

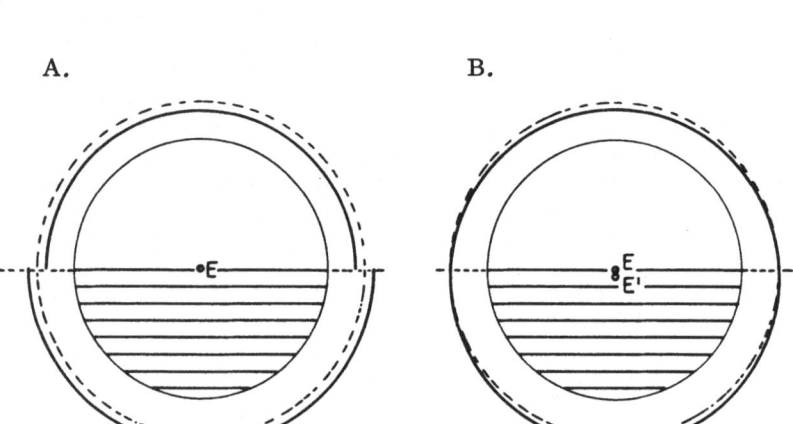

Fig. 17.1   a.   The lowering of the effective height of the
                 ionosphere on the side of the earth nearer to the sun.

             b.   The representation of this effect by a spherical
                  shell situated eccentrically with respect to the
                  centre of the earth.

In both cases, the total energy in the ring will have changed by a negligible amount so that the energy density on the surface of the earth will be inversely proportional to the radius of the ring i.e. $\propto 1/\mu$ as before. Thus, in the idealized case of a spherical smooth concentric earth and ionosphere it is found that, at the antipodal point (where $\mu = 0$) the power density becomes very great so that the antipode can be regarded as a point focus.

Using this as a standard of reference, we are led to enquire as to the type of distribution which would be expected in practice. First consider the effects arising from the fact that the ionospheric reflecting surface is not spherical and concentric with the earth as required in the idealization. It is difficult to represent the actual shape of the ionospheric reflecting surface by a simple mathematical expression. However, some general considerations lead to an approach to a solution. Let us choose some spherical surface, which is in fact concentric with the earth, as a "reference" surface. This surface could be, for example, the *average* virtual or effective height of reflection of rays at some specified frequency and angle of incidence (the average being taken over the whole earth over a 24 hour period), the exact specification of this reference sphere being unimportant. Then, during the night-time, the reflection height for a specified ray will generally be greater than the reference height and, during the daytime, because of the increased low-level ionisation which is present (in particular the F_1 layer) the reflection height for a particular ray will generally be lower than this reference height (fig. 17.1a). We will choose, the, as an approximation to the actual situation, the model shown in fig. 17.1b, where the reference ionospheric sphere has been shifted away from the sun so that its centre is at E'. Under these circumstances, the antipodal point loses much of its significance. In fact, the point focus which previously existed now becomes, in most cases, a caustic curve of the type usually observed in optics when reflections from cylindrical or spherical surfaces are studied. On the basis of this simple model of power distribution in the vicinity of the geometrical antipodal point can now be calculated, a sample distribution being as shown in fig. 17.3*.

---

* This distribution was calculated by Mr G.E.J. Bold.

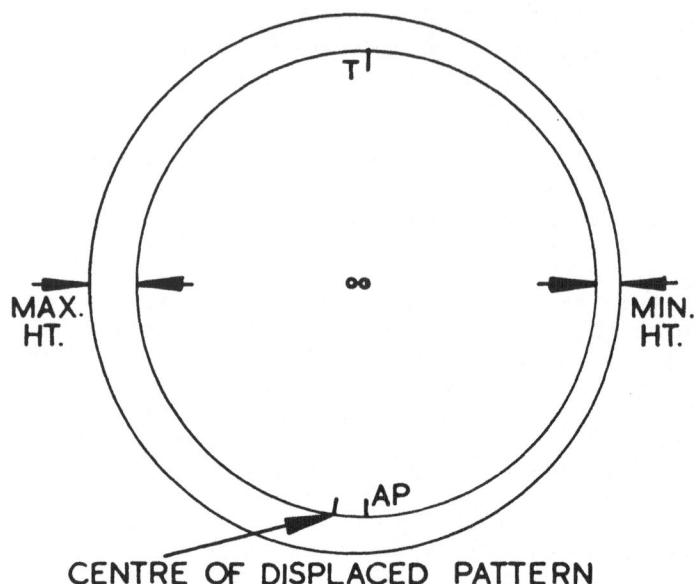

CENTRE OF DISPLACED PATTERN

*Fig. 17.2   The displacement of the centre of the focussing
region shown in fig. 17.3 by the eccentricity of
the ionosphere. For the case considered the
maximum height is 200 kms and the minimum height
is 150 kms.*

        In the reference case, since each contribution to the
power distribution from angles   very close to a particular
elevation angle at the transmitter was of the same $1/\mu$
form, it was unnecessary to consider the effects of the
vertical power distribution at the transmitting antenna.
However, if the eccentricity of the ionosphere is to be
introduced, these effects could be of importance. This
requirement introduces, in practice, the need that each
ray path be computed and that a divergence factor to account
for the inverse distance spreading of power be introduced.
Because of this latter effect and the general tendency of
long-distance transmitting antennas to concentrate the
power at relatively low angles, it is considered that
restricting the computations to elevation angles below
about 40° introduces negligible error.

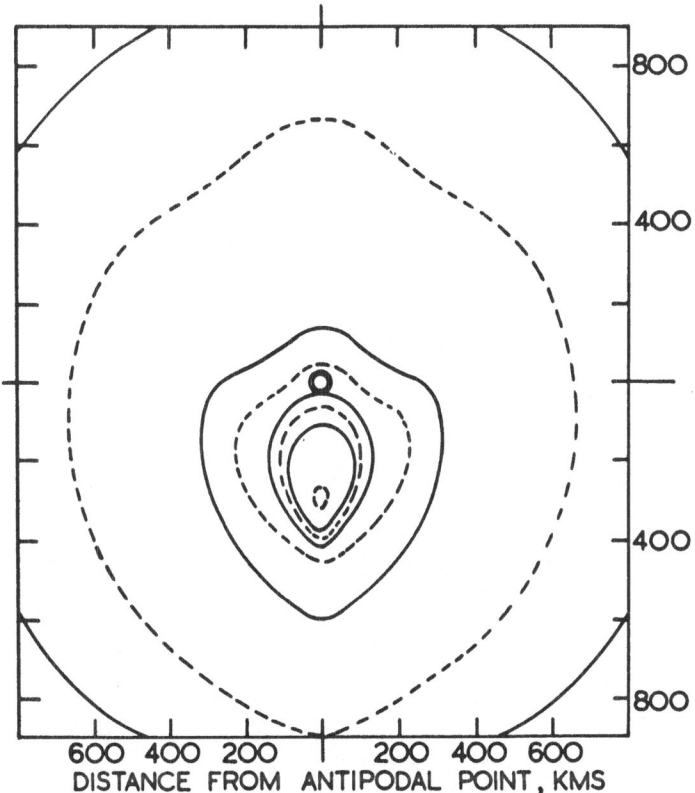

Fig. 17.3 Calculated distribution of power near the antipodal
point for the case illustrated in fig. 17.2

   The most important results which are obtained from these
computations are that some degree of antipodal focussing still
occurs but with distortion of the shape of the focussing area
when the ionosphere is situated eccentrically about the earth.
Associated with the distortion of the focussing area from
circular into a somewhat crescent-shaped region is a shift
of the whole area by an amount and in a direction depending
on the time of day. The direction in which the centre of
the area shifts is indicated in fig. 17.2. These results
have been obtained assuming a smooth earth and ionosphere;
in the next section the effect of roughness at the reflecting
points will be considered but, in order not to complicate the
situation too much, only for the case where the ionosphere
is concentric with the earth.

CHAPTER 18

ANGULAR SPREAD OF WAVES

The next step is to make some allowance for the fact that the reflections from the earth and from the ionosphere are not specular. The case which we are now considering is a little different from that considered in chapter 7 where the rough ionospheric and earth reflections gave rise to the day-to-day wandering of the rays. We are now concerned with the "instantaneous" picture i.e. the actual spread of power at some particular time.

In the treatment of chapter 7, a wave in a particular incident direction at one of the reflecting regions was regarded as giving rise to waves with a normal probability distribution of emergent directions. In the case to be considered her, we are concerned with the actual distribution of power (i.e. the angular power spectrum) in the ensemble of waves arising from each incident plane wave. We have already found that this angular power spectrum can not be represented by a simple normal distribution but contains two components *viz.* a pseudo-specular component and a scattered component. The latter is itself made up of a group of waves clustered about a mean direction which is commonly the direction of the pseudo-specular component. When this ensemble of rays meets another scattering screen, *each* of its components can be taken by itself and regarded as an individual plane wave. Each of these elementary plane waves will, after passing through the screen, become a plane wave of smaller amplitude (its own "specular" component) together with a group of scattered waves. All of these individual "specular" and scattered components must be added together to give the resultant angular power spectrum. This means that the number of terms to be considered in specifying the resultant angular power spectrum doubles at each reflection point so that the complexity of the problem rapidly becomes prohibitive. Fortunately, it can be shown (WHALE, 1966) that many of the terms are negligible and need not be considered.

The whole problem can be reduced to one which is quite tractable if we limit ourselves to those components which have been scattered once only. With this restriction, we have the situation where a ray originating at the transmitter travels on a great circle path losing a fixed percentage of

149

its energy at each encounter, this energy appearing as a
scattered component and travelling (again on a great circle
path) to the receiver. We must include the further loss of
energy from this scattered component each time it meets the
scattering reflecting region, the fixed percentage loss of
energy being the same as that applicable to the specular
or plane wave component but can ignore the question of the
fate of this energy which is "lost" (further scattered)
from these scattered components since the total energy
involved will generally be quite small. The ensemble of
waves which finally arrives at the observing point is thus
made up of a small specular component together with a number
of scattered groups of waves of different spreads. In this
analysis it is assumed that all reflection points have the
same average properties. The fraction of energy that is not
scattered at a reflection point is the specular component
and this is specified in terms of the coherence ratio B.
Then, if the coherence ratio is B_1 after one reflection of
a plane wave, it will be given approximately by

$$B = \left(\frac{B_1}{1 + B_1}\right)^N \text{ after N reflections.}$$

From the relatively few measurements made of B at
various distances from a transmitter we find that B_1 has an
average value of about 3 on making some allowance for the
fact that B is not constant. This indicates that after
about 20 reflections (10 hops) the coherence ratio will
be quite small. The other factor which is needed to specify
the distribution of waves in the incoming angular power
spectrum at any distance from the transmitter is some measure
of the width of the scatterd part. The total power in the
scattered part is, of course, specified by B obtained from
the above. We have also noted that the scattered part consists of a large number of scattered groups of waves which
have travelled different distances over different paths.
These groups will have spreads which are different because,
although the groups of waves are all assumed to have arisen
by the same type of scattering mechanism at the different
reflection points, these reflection points are situated at
a series of different locations with respect to the transmitter and the receiver, which is sufficient condition for
the spreads observed at the receiver to be different. One
interesting result is that the amount of energy finally
contained in each of the spread components is the same.
This arises since both the plane wave and the scattered wave

lose a fixed percentage of their power at each reflection,
and since each received wave will have travelled part of
the distance as a plane wave and the rest as a scattered
wave, the total number of reflections and hence the total
percentage loss of energy is the same for all rays. Thus
the total angular spectrum of waves observed at the
receiver will be made up of a series of constituent components, all of gaussian form and, since they contain the
same energy, of the same area with the widely spread components having a smaller maximum value than the sharply-beamed
components.

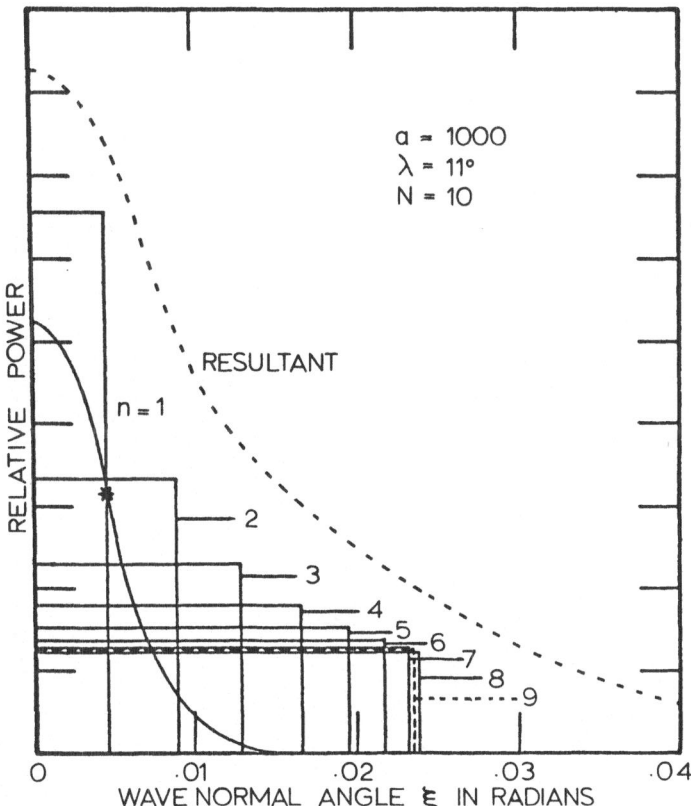

*Fig. 18.1 The way in which the incoming fan or rays is made
up of a number of component distributions. These latter distributions are represented by rectangles with the correct
area and with width equal to the standard deviation.*

A typical ensemble of these waves is drawn in fig. 18.1.
The overall spread is drawn as a dotted curve in the same
diagram. It is noted that this is a much more sharply peaked distribution than the normal or gaussian type. In statistical theory, this type of distribution would be said to
be leptokurtic i.e. more sharply peaked than a normal distribution. Both the second moment ξ^2, which describes the
width of the distribution and the fourth moment ξ^4 may be
calculated. These moments can be used to indicate the peakiness or kurtosis K of the distribution which may be expressed in the form which was given previously in equation
14.6.

The value of K for a normal distribution is zero. The
distributions which are obtained here are rather more peaked

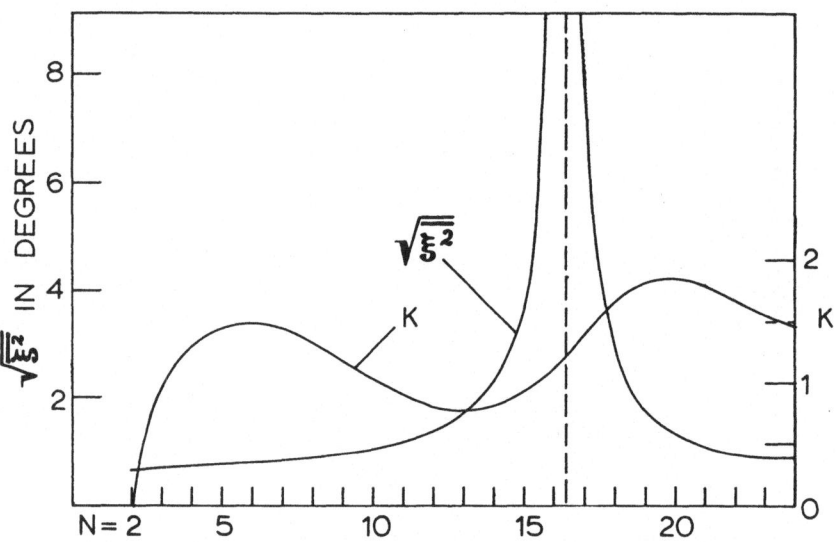

Fig. 18.2 *The variation of the spread of the scattered part
of the incoming distribution and the kurtosis as
a function of number of reflections.*

than normal curves so that K is a positive quantity. In
fact, K does not vary very much with distance. A computed
plot of the standard deviation of the spread of the scattered
part of the incoming fan of rays $\sqrt{\xi^2}$ and of K is drawn in
fig. 18.2. In order that these curves could be calculated,
it was necessary to make some assumptions about the characteristics of the scattering process at each reflection point.
It will be remembered that, when considering the day-to-day
deviations of the bearing, a scattering factor A which described the width of the probability distribution of scattered
directions at an average reflection point was introduced and
it was found that A had a typical value of about 1000. In
the present case two parameters are required in order to
describe the scattering *viz.* a factor defining the coherence
ratio of the scattered part of this wave system. Assume
that the scattered part obtained when a plane wave has suffered one average reflection is given by the gaussian form

$$p(\xi) \propto \exp - a\xi^2$$

The **values** of $\overline{\xi^2}$ given in fig. 18.2 are then those obtained for the *scattered part only* when a = 1000. We find
that the value of K does not, in its final form, depend on
the value assumed for a. Again, this curve for K is for the
scattered part only of the incoming group of waves since the
final coherence ratio is so small that the specular component is negligible.

It may be a little difficult to check the results for K
experimentally since most measurement methods which employ
direction-finders give a value for $\overline{\xi^2}$ but do not yield the
higher moments directly. For example, consider the type of
measurement which yielded the curves drawn in fig. 14.3.
The horizontal spread of these distributions depends on $\overline{\xi^2}$.
If there is an appreciable third moment ξ^3 the shapes of the
distributions are not altered significantly although the
whole distribution is displaced sideways by a small angle.
An appreciable fourth moment ξ^4 simply has the effect of
altering the apparent value of $\overline{\xi^2}$ so that these experimental
distributions are not useful for determining the shape of
the incoming distribution of rays i.e. the angular power
spectrum.

The shape of this angular power spectrum can be determined only if the shape of the generalized auto-correlation
function across the ground is known since the two distributions are related (each being the Fourier transform of the
other). We know $\rho(0) = 1$ i.e. the value of the generalized
auto-correlation function at zero antenna separation is, by
definition, unity. We next find the value at very large
separation and obtain from it a value of B. A third value
at some intermediate point will provide a measure of the
second moment ξ^2 of the incoming distribution. A further
measurement at another distance will provide information on
$\overline{\xi^4}$, etc. We note that we have assumed a symmetrical distribution in obtaining the curves in fig. 16.3 which are the
basis of this type of measurement so that all the odd moments
of the incoming distribution must be assumed to be zero.
The presence of odd moments (asymmetries) in the distribution,
in fact, leads to periodicities in the generalized auto-
correlation function curve.

No systematic investigation of the shape of the incoming
angular power spectrum and its variation with distance has
been carried out. Some measurements of the way in which the
spread of the incoming fan of rays varied with distance have
been attempted in connection with the obtaining of distribution
like those in fig. 14.3. The experimental procedure is difficult because the frequency of the scintillations tends to
increase as the propagation distance increases. But the
incoming spectrum is made up of many groups of waves which
have travelled such different distances so that, if there
is a limit to the rate of change of the scintillations which
the equipment can follow, some of these component groups
may have little effect on the recorded directional variations.
The equipment which was used to obtain the joint probability
distributions exemplified by fig. 14.9 was one in which an
antenna system locked on to and followed the fluctuations
in wave-normal direction. There was thus an appreciable
time-constant involved so that the more rapid parts of the
fluctuations could not be recorded. This is an effect which
is in addition to the loss of the tails of the distributions
which has already been discussed in section 14.

When the above factors are taken into account it is
found that the measured results do, in fact, agree quite
well with the theoretical curve for $\sqrt{\xi^2}$ in fig. 18.2. These
results are plotted in fig. 18.3 in which the spread (standard deviation) of the fluctuations in wave-normal direction

is plotted as the vertical lines as a function of distance
between the transmitter and the receiver. This spread
is, of course, a quantity which is different from the spread
of the incoming fan of rays although the two are related
in the way which has been discussed previously. In obtaining the three sets of curves in this figure, the following
factors have been considered. Firstly, the dotted curves
are taken directly from the curves for the spread such as
that in fig. 18.2. These curves make no allowance for the
presence of any specular component. In this case we have
used the relation which we obtained previously as equation
14.6 and which is valid when the spreads are small, that

$$\overline{S^2} = 1 \cdot 2 \overline{\xi^2}$$

Secondly, the dashed curves are obtained from these by
making reasonable assumptions about the frequency response
of the measuring equipment. In fact, it has been assumed
that the equipment response follows a 1/f law (where f is
the frequency of the fluctuations). For this particular
law, if it is also assumed that all the reflecting regions
are similar in that the same rates of change occur at each,
it is possible to derive the shape of the *measured* distribution of incoming wave-normal directions. For each of
these cases, a few representative values of the scattering
parameter a have been chosen.

Thirdly, the full-line curves have been derived from
the latter curves by including the effect of a specular
component. In this case it has been necessary to estimate
the value of the coherence ratio for an average single
reflection from the earth or the ionosphere. It has been
mentioned previously that, from the measurements which
have been made, it would appear that B = 3 is a reasonable
average value. Accordingly, this value has been used in
deriving the curves in fig. 18.3. It can not be said that
these follow the measured values (indicated by the vertical
flagged lines) very closely. A more detailed investigation
of the problem is obviously required. The one important
result which can be obtained from these curves is an
estimate of the average value of the scattering parameter
a and it is seen that this is approximately 300.

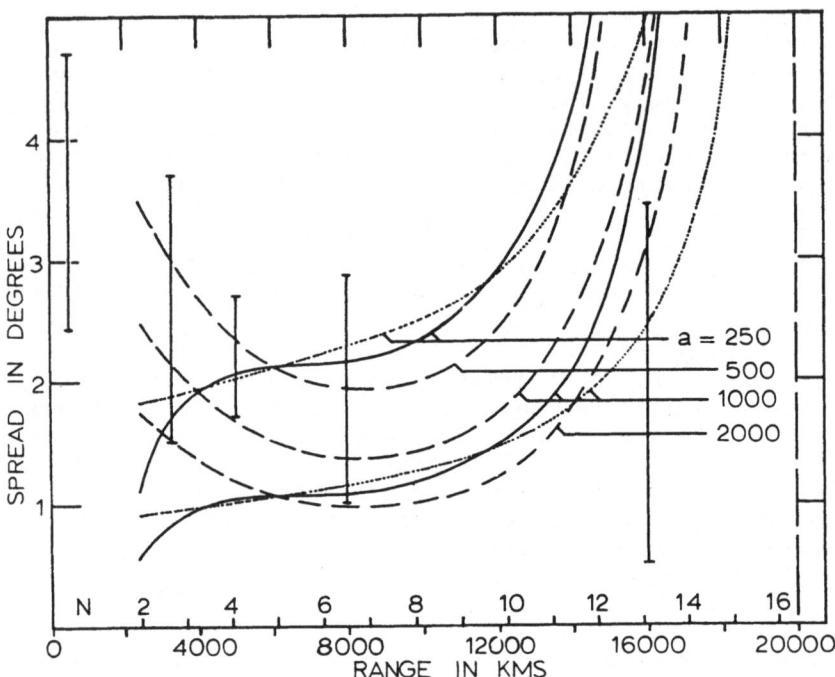

Fig. 18.3 The various factors to be considered in relating
the theory to the experimentally obtained spreads.

CHAPTER 19

THE ANTIPODAL AREA WITH A ROUGH OR
ABSORBING IONOSPHERE

Since a percentage of the energy in the specular component is removed at each reflection there will always be some
specular or pseudo-specular component left. This remaining
energy has not been scattered and thus it will always travel
on a great circle path and will all pass through the point
antipodal to the transmitter; the idealized type of antipodal
focussing which was discussed in the first part of chapter 17
will occur there. In any practical case, the departure of
the ionosphere from the ideal smooth spherical shape will
cause the focussing of the specular energy to be slightly
"blurred". The scattered energy arising at each reflecting
point will be "beamed" towards the point antipodal to the
transmitter but will not be focussed there since the angular
distance from the point where the scattered energy originates
to the point which is the transmitter's antipode is always
less than 180°. There will, however, still be a general
clustering of the scattered energy in the vicinity of the
point which is antipodal to the transmitter.

This process leads to two effects, either of which may
be used as a basis for defining the extent of the antipodal
area. Firstly, signals will be arriving over a wide range of
angles as has already been pointed out in connection with
fig. 18.2. If the standard deviation of the spread of the
incoming group of waves is given by σ, then it is found that,
near the antipodal point,

$$\sigma = \cdot 5/(\mu\sqrt{a}) \text{ radians.} \qquad \ldots (19.1)$$

In this expression, μ is the angular distance of an observing point measured in radians from the antipodal point
and a is the scattering parameter which describes the spread
of the scattered waves obtained from a plane wave at any reflection point.

Secondly, the antipodal area may be specified in terms
of the distribution of the intensity of the radio energy in
its vicinity. This intensity or power distribution, when
plotted as a function of μ (defined above) is rather similar
to a gaussian curve but is, like the various other angular

157

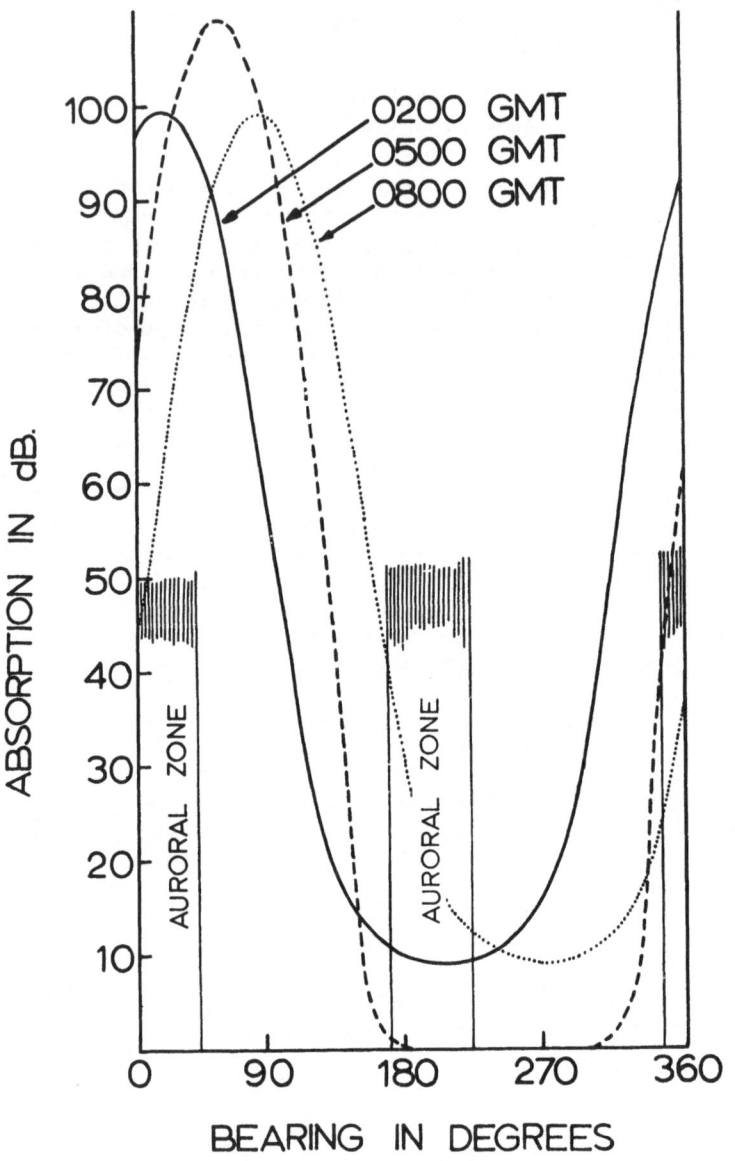

Fig. 19.1. Absorption along a great circle path from Tangier to Auckland, originating along the bearings shown, at a frequency of 11790 kHz for the month of May when the relative sunspot number is 100.

distributions which have been considered in the last few
chapters, rather more sharply peaked. If, for reasonable
values of the scattering parameter a the antipodal area is
defined as the zone within which the power density is greater
than half its central (maximum) value it is found that its
radius is given in angular measure by

$$\mu = \cdot 5/\sqrt{a} \qquad \qquad \ldots \ldots (19.2)$$

Note that this agrees with the former expression (19.1) for
the spread of waves if the antipodal area is defined as that
zone within which the standard deviation (width) of the incoming fan of rays is greater than 1 radian. In general, the
focussing becomes less pronounced as the amount of scattering
at each reflection point and the number of hops increases.

The changes produced because the ionosphere is neither
spherical nor concentric with the earth are of significant
importance since these generally appear as a broadening of
the antipodal area, a distortion of its shape and perhaps a
displacement of its position. The experimental results required to give a good measure of the magnitude of the scattering parameters (B and a) which should be used in the above
development are of several kinds. Some of the approaches
which have been employed are:

(i) the direct measurement of the spread of the incoming cone of rays at various distances from a transmitter,

(ii) the measurement of bearing angle and elevation
angle of the various sub-groups of the group of pulses which
is received at a distant station when single pulses are
transmitted,

(iii) the direct measurement of the size and shape of
the antipodal area by the taking of many fieldstrength measurements over a large area.

None of these methods has been fully explored although
some preliminary measurements have been made in each case.

Only a few experimental results are available for comparison with equation 19.2. The direct measurement of the
size of the antipodal area has been attempted in one experiment (PIPP and WEBSTER, 1964). These measurements were obtained by carrying out aeroplane flights in the vicinity of

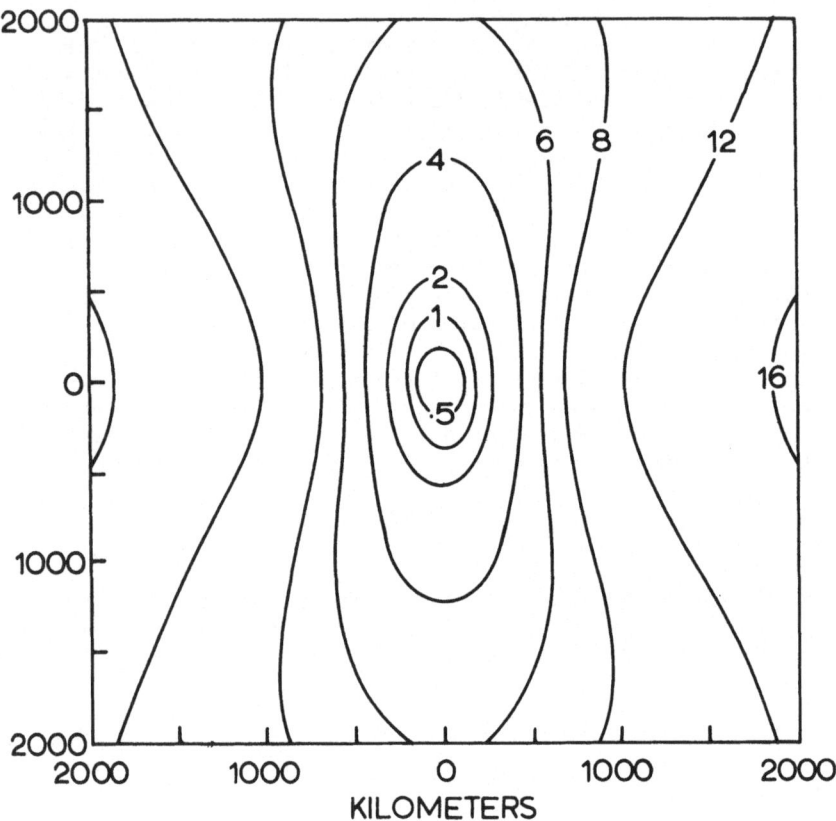

Fig. 19.2. Contours of equal power near the antipodal area for a beamed transmitter; beam-width = 60°, a = 3000.

the geometrical antipodal point. There was considerable difficulty in removing the normal diurnal changes in received power from the results but the indications are that a measured radius of about 300 kilometres could be assigned to the antipodal area.

Expression 19.2 specified the width of the antipodal area in which focussing occurs without specifying the actual shape of the distribution. It is possible to specify this shape by considering the total power which comes from all directions in which case measurements of the relative power received at two points in the antipodal area should give enough information to specify the curve. A relevant measurement was carried out by GERSON, NARDOZZA and HENGEN (1962) when they found the relative percentage of time that a 16 MHz signal transmitted from Perth, Australia was received at two locations, one of which was the antipodal point (Bermuda) and the other of which (Rome, N.Y.), was 13.5° angular distance away from the antipode. The results were presented as the number of times the signal was received during each hour of the day over a six weeks period. The total number of times of reception for each station will represent the relative average power received at that station and can be compared with the theoretical power distribution in the antipodal area. It was found (WHALE, 1963) that this ratio was approximately .5 and agreed reasonably well with the assumption that the scattering factor a was about 1000. No allowance was made for several factors which could be important. Firstly, it is known that the antipodal area moves in position; secondly, it is also known that not all paths between the transmitter and receiver will simultaneously allow propagation to occur. Nevertheless, regarding the numbers obtained as an overall averaged result, it is likely that they have some significance.

A further factor which must be considered is that all great circle paths between the transmitter and a receiver near its antipode are not available or of equal importance. The different amounts of power received from different directions will depend on the absorption over different paths (in the case of an omni-directional transmitting antenna) and also on the antenna beaming direction (in the case of a directional transmitter).

As well as the effects arising from absorption in the auroral regions, there are those arising from the normal

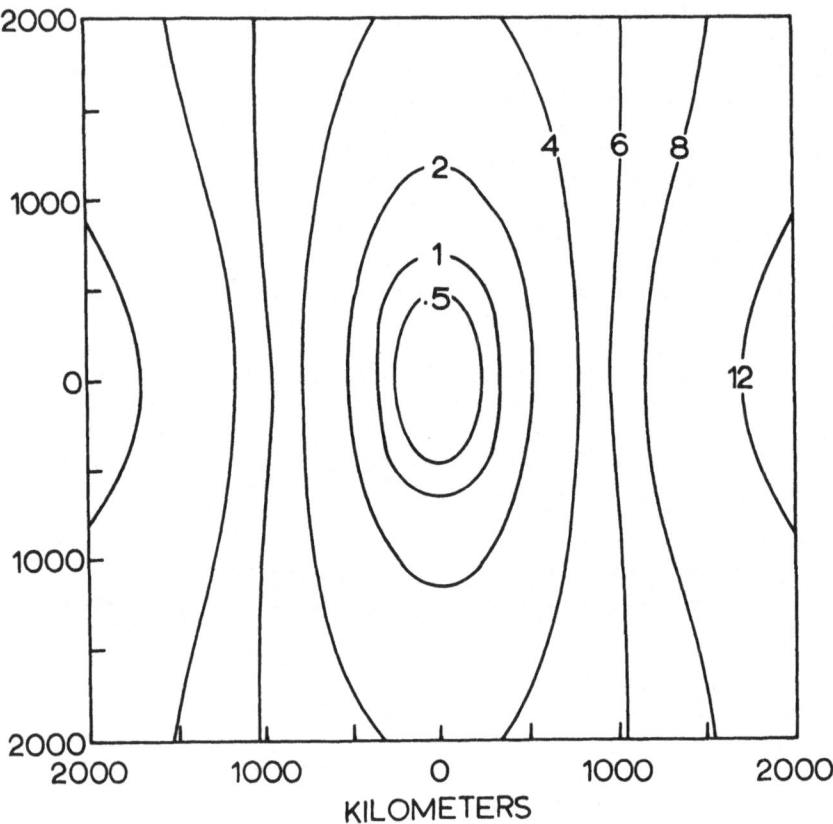

*Fig. 19.3. Contours of equal power near the antipodal area
for a beamed transmitter; beam-width = 60°, a = 1000.*

enhanced absorption which occurs primarily in the sunlit
hemisphere. This latter will be symmetrical only when the
sun is vertically overhead at the transmitter or its antipode.
At all other times there will be some preferred path over
which the absorption is least. Since this path will pass
through or close to the antipode of the subsolar point its
bearing (measured at the transmitter, say) will change with
the time of day. As an example of this behaviour, the total
absorption along a great circle path from a transmitter on
11790 kHz located at Tangier to its antipode (at Auckland)
has been calculated for various bearing directions measured
at the transmitter. These curves are plotted in fig. 19.1
for three different times of day for the month of May, assuming a sunspot number of 100. When the extra, usually
large, absorption arising in the auroral zones is included,
it can be seen that, for the time of day considered (0200 to
0800 GMT) the most favourable paths are concentrated near
the 260-270 degree bearing angles.

Much the same effect would be obtained if there were no
absorption but the energy radiated by the transmitter were
confined to a small range of angles i.e. a beamed transmitting
antenna were used. As examples, the contours in figures 19.2
and 19.3 have been computed. These are contours of equal
power density near the antipode of a transmitter with total
beam-width 60° (as specified in figure 7.4) oriented so that
the direction of maximum incoming radiation is along a line
that is in the ordinate direction in the figures. The contours are labelled in dB below the value of power density at
the antipode where the maximum occurs. In fig. 19.2, a
scattering factor (a) of 3000 has been assumed; this implies
a rather small amount of scattering at each reflection point.
About 20 steps have been assumed to occur between the transmitter and its antipode in all cases. In fig. 19.3, a rather
larger degree of scattering (a = 1000) has been postulated.
The general effect is a broadening of the maximum while the
ellipticity has remained about the same.

In fig. 19.4, the same scattering factor (a = 1000) has
been used, but the antenna beam-width has been reduced to 25°.
The effect has been to increase the ellipticity of the antipodal area considerably from that obtained with the 60° beam-
width in fig. 19.3.

Some recent experimental results obtained by Mr G.E.J.
BOLD, who also computed the above curves, indicate that the

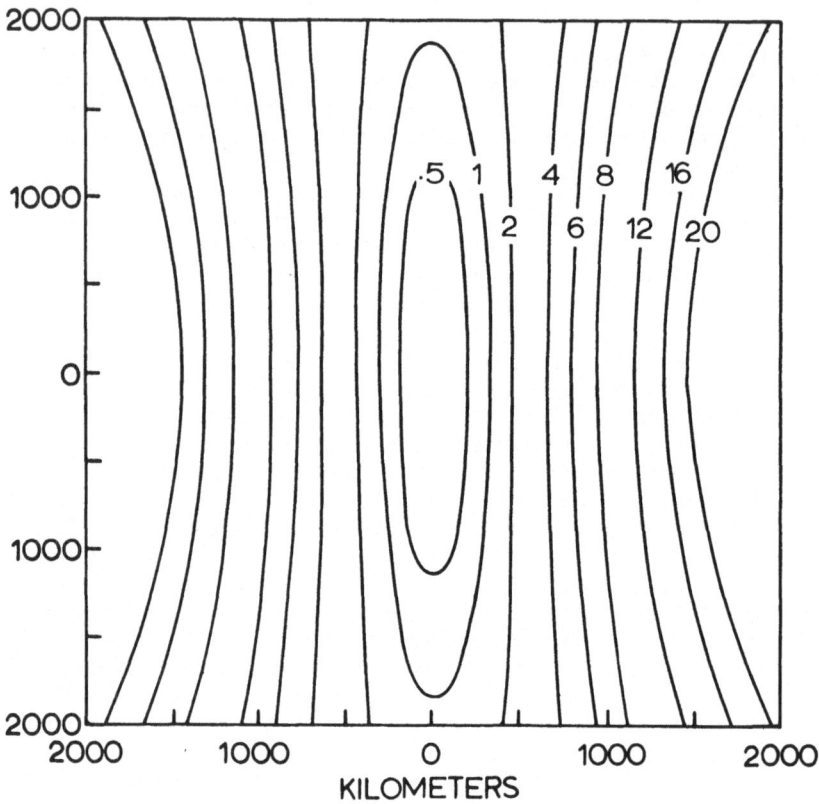

Fig. 19.4. Contours of equal power near the antipodal area for a beamed transmitter; beam-width = 25°, a = 1000.

antipodal area existing when a directional transmitting antenna is used is indeed of this elliptical shape.

It is relevant to enquire, at this stage, into the effects of the scattering in the vertical plane on the deductions made above. Any such scattered wave will still follow a great circle path after it has been generated; the step lengths will be different from the original step length because of the change in elevation angle during the scattering process but the energy will still flow across the antipodal region in a beam whose width is determined by the horizontal scattering parameter (a) and the distance which the beam has travelled. This leads to the general conclusion that the vertical scattering is relatively unimportant in the above work and may safely be ignored.

CHAPTER 20

Duct Propagation

There is considerable evidence that, under some
circumstances, waves may propagate over long distances
without returning to earth at periodic intervals as they
do in hop-mode propagation. This means that the wave
travels either in a series of hops between two ionospheric
layers (perhaps the E and F layers) or as a ducted wave
which is constrained by the refractive index profile of
the ionosphere to remain at about the same level in the
ionosphere. In both cases, any theoretical approach must
explain the mechanism by which the energy enters and leaves
the region where this type of propagation can occur.

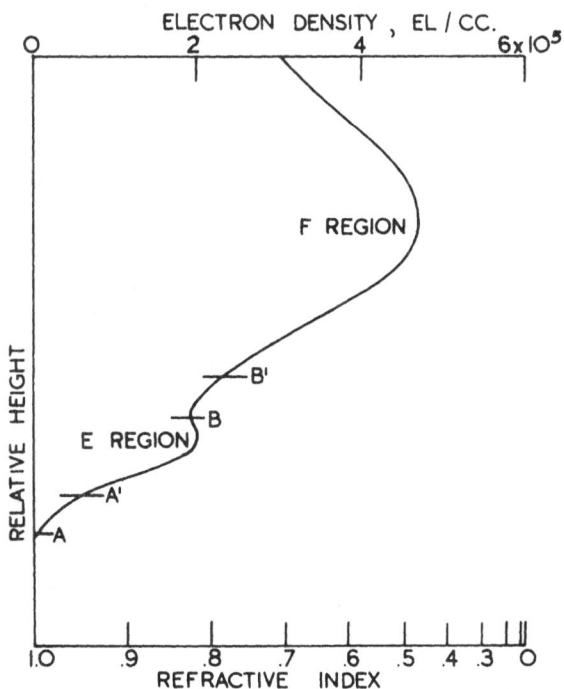

Fig. 20.1 Representative ionospheric profile showing the
regions favourable to duct-type propagation.
Typical values of electron density and refractive
index are given in the top and bottom scales respectively.

Of particular interest in this connection is the case
of long distance propagation from a satellite which is
transmitting in the short wave band. Cases of reception at
very long distances were particularly noticeable with
Sputnik 1. In such a case, the transmitter will be situated either in or very close to any ducting regions which
may exist so that the problem of how the energy enters the
region does not occur. There is still, however, the problem
of how it leaves the region. In the following, we will
discuss the way in which ducts occur and give some thought
to the problem of feeding energy into and out of such a
duct.

A typical daytime ionospheric profile would be like
the one plotted in fig. 20.1. This is a plot of electron
density plotted horizontally (the scale is at the top of
the graph) as a function of height. For each electron
density, there corresponds, for a particular frequency of
wave, the *refractive index* given by

$$\mu = \sqrt{1 - KN}$$

where the constant K depends on the frequency being propagated and N is the electron density. From this relation, fig.
20.1 can also be regarded as a plot of refractive index as
a function of height (the refractive index scale is give at
the bottom of the graph). A similar problem was investigated in the early days of radar when a very exhaustive
study of the effects of the shape of the vertical refractive
index profile of the lower atmosphere on the propagation of
radio waves was carried out in connexion with the phenomenon
that was then called anomalous propagation. In that case
the variations in refractive index arose from variations in
the humidity and temperature of the atmosphere at low levels
i.e. near ground level. In the ionosphere, as in any other
medium, the wave velocity (or phase velocity) increases
as the refractive index decreases. If a region exists where
the decrease of refractive index with increasing height is
of the correct value, then ducting can occur. The requirement for the correct value of this slope (rate of decrease
with height) of the refractive index profile can be seen
if we consider two cases. Firstly, if the refractive index
were constant, the waves would travel in straight lines
with no tendency to curve around the earth as in duct type
propagation. Secondly, if the refractive index gradient
were too steep, the wave would be bent sharply and thus be
reflected off this region. There is some intermediate slope

which is just sufficient to introduce the correct curvature
into a horizontal ray so that it will continue around the
earth at about the same level. There is a small range of
angles near this level which will also be effectively
trapped in the same way since rays which are travelling
slightly upwards (as in fig. 20.2) will meet lower refractive indices and will be bent downwards so that they travel
back through the ducting level and, if the refractive index
variation is small at these lower levels, such rays will
travel in almost straight lines and will thus encounter
the upper region again after traversing a chord BC. Rays
which start by travelling in a direction which is slightly
downwards may be considered as starting at B, traversing the
chord BC and entering the region of decreasing refractive
index at C. The latter mode is sometimes referred to as
a chordal hop or whispering gallery mode; both this mode
and the one discussed above in which a continuous bending
of the ray by just the right gradient of refractive index
occurs have been suggested as mechanisms explaining some
types of ling distance propagation. The proper gradients
would occur in the two regions AA' and BB' in fig. 20.1.
The region AA' is very near the base of the ionosphere
where the atmospheric density is relatively high. Under
this condition, the radio wave suffers a loss of energy
since it is in this low region of the ionosphere that most
of the absorption of waves that is important in long-distance communications occurs. A wave which is trapped so
that it remains in this region will rapidly become completely absorbed so that region AA' need not be considered
further. There is no such objection to propagation in
the region BB' so that this must be considered as a possible
ducting region.

Fig. 20.2 Chordal hop type propagation.

It is in fact the presence of these absorbing regions in the lower ionosphere which leads to the supposition that ducting is occurring. Some of the experimental evidence arises from a study of signals which have travelled several times around the world. For this type of investigation it is simplest to use pulse transmissions which are short enough so that each round-the-world signal is separated in time when it arrives at the receiver. If each circuit of the world is of the same nature i.e. the wave travels from the transmitter to the ionosphere and then round the world in a series of hops, the loss of signal strength by absorption will be roughly equal for each complete circuit. In some circumstances signals which have travelled several times around the world have been observed. The loss in strength between successive signals is easily measured and has been found to be of the order of 10 db. The loss suffered by the first received pulse in its passage from the transmitter to the receiver is not as easily found since this involves comparing the field-strength of the received signal with that deduced from a knowledge of the power radiated by the transmitting antenna in the direction corresponding to that taken by the ray. The loss so calculated appears to be of the order of 90 db. Quite obviously, the signals which have travelled more than once around the world have followed a path which is quite different from that followed by the signal first received. Of the 90 db path loss which the first signal suffers, about half of the 80 db excess will arise at the transmitter end in establishing the low-loss mode and the other half at the receiving end in getting energy out of this mode. It is interesting that, if the signal were established within the ionosphere so that the loss at the transmitter end were avoided, the signal at the receiver should increase by about 40 db. Now the Sputnik 1 transmitter was about 1 watt in power and gave signals at the ground near its antipodal point which were approximately the strength which would be expected from a ground station radiating a few kilowatts. Ten kilowatts is 40 db above 1 watt.

The problem of the way in which energy enters and leaves an ionospheric duct has received some attention; the general solution which has been offered requires the presence of substantial tilts in the upper ionosphere at both the transmitting and receiving ends of the path so that the energy is guided into and out of the duct. One consequence

of this theory is that such a mechanism, while favouring a
single traverse of the ionosphere by a ducted mode, hardly
explains the observed fact that several circuits of the
earth seem to be possible.

The theory of ducting which has been discussed in
connection with figures 20.1 and 20.2 is essentially one
which assumes that rays propagate in the ionosphere in a
manner predictable from the assumption of a smoothly varying
medium. If only a short path in the ionosphere is involved,
this assumption may be valid but it appears unlikely that
such modes would be supported by the ionosphere over large
distances. A more realistic approach would appear to be
one which is based on the scattering properties of the
ionosphere.

We have already seen that the roughness of the ionosphere leads to considerable deviations of the bearing
directions of waves received at a distant point and, in
fact, makes possible the reception of signals in areas
which would be regarded as being shadowed by obstacles such
as the auroral absorbing regions. We have also seen, from
fig. 13.3 that these irregularities are distributed throughout the ionosphere so that they will affect the vertical angle
in a way which is similar to that for the horizontal angle.
Those regions of the ionosphere which appear to be absorbing
or opaque to waves at various special angles of elevation
may be regarded as "stops" in the same way as the auroral
region was previously regarded as a stop and it then becomes
clear that it is possible, in many cases, for waves which
have been scattered from different irregularities at different places to avoid such regions and finally arrive at
the receiver.

The experimental investigation of this type of propagation which could, in fact, be of the utmost importance
in maintaining long-distance communications, is rather
difficult. One of the important considerations arises from
the fact that waves arriving by a normal hop mode should
contain at least a small specular or pseudo-specular component while waves which have been scattered should contain
no such specular component. We have already seen, however,
that the specular component is likely to be very small
in cases of long-distance hop type propagation but it is
these long-distance paths which are most likely to exhibit
effects arising from the presence of duct modes. It is for

this reason that a considerable effort is being made at the
present time towards developing the accurate methods of
measurement of the coherence ratio which have been described
in chapter 16.

CHAPTER 21

CONCLUSION

Although very little reference has been made in this text to the classical approach to the problems of ionospheric propagation, that great volume of work has formed the basis of all that has been described here. If all the relevant characteristics of the ionosphere are known for any particular time, then a process of ray-tracing or of the application of wave-theory to the problem must yield the correct answers. Such an approach has, in fact, proved extremely valuable in the investigation of short to medium distance propagation paths. Very sophisticated instruments such as oblique incidence ionospheric sounders have been developed and employed to demonstrate that the mechanism of ionospheric propagation is in fact very well understood and is amenable to mathematical treatment. As the distances increase however, so does the number of unknown and usually random factors. At the very great distances which have been considered here the random factors outweigh the regular behaviour of the ionosphere to the extent that it does not appear to be fruitful to attempt to tackle the problem by methods which are extensions of the short-distance techniques.

The ionosphere as a whole is such a changeable medium that a very large number of measurements is usually required before any facet of its behaviour can be described adequately, let alone predicted with any certainty. It was most fortunate that an instrument which could be used to provide a very large number of accurate measurements of the angle-of-arrival of radio signals was developed early in the project. Rotating interferometers, because of their continuous recording capabilities, made possible the acquisition of enough data so that at least what the problems were that needed to be solved could be specified.

It must be emphasized that very few of the subjects treated in this work have been studied fully and it is probable that many of them never will be. It is known that the greater portion of the power radiated by a transmitter is wasted and never reaches the vicinity of the desired receiving installation. Although the great majority of this work has been concerned with the reception of that small portion

of energy which does, in fact,find its way to the receiving
antenna one is intrigued by the possibility that the same
results could be achieved with a tremendous saving of effort
if the energy from the transmitter were only sent in the
proper direction in the first place. It has been noticed,
for example that signals from European stations seldom
completely disappear even when they would be considered quite
useless for the transmission of intelligible information.
The rotating interferometers, with their inherent very great
sensitivity because of the phase-switching technique employed,
continue making reliable measurements of the bearing and
elevation angle of signals which have become so weak as to
be barely detectable by ordinary standards. As an example
of a possible line of approach, it is felt that a better
understanding of the mechanism of duct propagation, a subject
which has been treated only briefly to date, could lead to
substantial improvements in long-distance communications.

Major advances in the technology of communicating over
long distances have been made with the commissioning of the
communications satellites. These potentially could lead to
the achievement of the ultimate aim in communications *viz.*
the possibility that any person in the world could communicate rapidly and reliably with any other person. The bene-
fits of this to world-wide understanding and co-operation
are incalculable but this ideal objective may take some
time to eventuate. Even though satellites have been most
generously made available to all countries there is still
a case to be made for improving those means of communication
which are not subject to any national control or ownership
and it is for this reason that the work described here is
continuing and that this text has been produced.

REFERENCES

Booker, H.G., J.A. Ratcliffe and D.H. Shinn, Diffraction from an irregular screen with applications to ionospheric problems, Phil. Trans. Roy. Society London, A242, 579-607 (1950)

Bramley, E.N., Some aspects of rapid directional fluctuations of short radio waves reflected at the ionosphere, Proc. Inst. Elec. Eng. 102B, 533-540 (1955)

Davies, K., Ionospheric Radio Propagation, National Bureau of Standards Monograph 80 (1965)

Gerson,N.C., V.J. Nardozza and J.G. Hengen, High frequency propagation over long paths, J.Geophys. Res. 67(10), 4084-4085 (1962).

Mitra, S.K., The Upper Atmosphere, The Asiatic Society, Calcutta (1947)

Pipp, R.M. and J.B. Webster, An experimental investigation of signal strength in the area around a transmitter's antipode, Radio Science 68D, 333-337 (1964)

Rice, S.O., Mathematical analysis of random noise, Bell System Technical Journal 23, 282-332 (1944).

ibid, 24, 46-156 (1954).

Titheridge, J.E., Variations in the direction of arrival of high-frequency radio waves, Journ. Atmosph. Terr. Phys. 13, 17-25 (1958).

Whale, H.A., Effective tilts of the ionosphere at places about 1000 km apart, Proc. Phys. Coc. 69B, 301-310 (1956)

Whale, H.A., The effects of ionospheric irregularities and the auroral zone on the bearings of short-wave radio signals, Journ. Atmosph. Terr. Physics 13, 258-270 (1959).

Whale, H.A. Ionospheric scattering effects in long-distance propagation, Journal of Research of NBS 67D, 287-296 (1963)

Whale, H.A., Comparison of experimental data with the theory of the antipodal area, J. Geophys. Res. 68(14), 4390-4391 (1963).

Whale, H.A. The angular spread of radio waves in long-distance propagation, Radio Science 1, 743-750 (1966).

Whale, H.A. and W.J. Ross, An automatic direction-finder for recording rapid fluctuations of the bearing of short radio waves, Proc. Phys. Soc. 69B, 311-320 (1956)

Whale, H.A. and R.W. Bannister, The rotating interferometer and its use in measuring the spread and coherence ratio of a scattered radio wave, Radio Science 1, 1237-1244 (1966)

This selection of references is not intended to be comprehensive but merely to act as a guide towards obtaining more detailed treatments of some of the subjects covered in the text.

INDEX